Vibration in Mechanical Systems

Vibration in Mechanical Systems

By

Cho Wing S. To and Qishao Lu

Cambridge
Scholars
Publishing

Vibration in Mechanical Systems

By Cho Wing S. To and Qishao Lu

This book first published 2023

Cambridge Scholars Publishing

Lady Stephenson Library, Newcastle upon Tyne, NE6 2PA, UK

British Library Cataloguing in Publication Data
A catalogue record for this book is available from the British Library

ISBN (10): 1-5275-3361-1
ISBN (13): 978-1-5275-3361-5

To Kem, Leighton, Jane and Anita

CONTENTS

PREFACE

This book is in the form of lecture notes and was originally created for two semesters of 6 credit hours courses taken by senior undergraduates and first year graduate students majoring in mechanical engineering. The proposed book initially consisted of 15 chapters. They are based on the lecture notes for a required course with similar titles given to junior (and occasionally senior) undergraduate students by the author in the Department of Mechanical Engineering at the University of Calgary from 1981 and at the University of Nebraska, Lincoln since 1996. The emphasis is on fundamental concepts, theory, analysis and design of mechanisms with applications. While it is aimed at senior undergraduates and first year graduate students majoring in mechanical engineering, it is also suitable for senior undergraduates and first year graduates in biological system engineering, ocean engineering, naval architectural engineering, and aerospace engineering.

The original framework of this book was intended to draft for four parts and fifteen chapters in 2018 and its proposed delivery date was October 30, 2020. The title of the proposed book would be: **Vibration, Dynamics, Stability and Bifurcation in Mechanical Systems**. The four parts included: **Part I. Vibration in Mechanical Systems, Part II. Balance, Dynamics in Rotors, Engines, and Machines, Part III. Contact Dynamics in Machinery, and Part IV. Stability and Bifurcation Analysis in Mechanical Systems.**

By February 2020. **Part I. Vibration in Mechanical Systems** had mostly been completed while the author was still active. Now the book is named after Part I. **Vibration in Mechanical Systems**. Owing to the absence of the original author, the author's excellent writing habit of careful and detailed styles might meet challenge in the production of this book.

The author's first book **Nonlinear Random Vibration** was published in 2000. **Vibration in Mechanical Systems** will become his 8th book to be published.

Cho W. S. To

ACKNOWLEDGEMENTS

Vibration in Mechanical Systems was intended to write up the author's knowledge and experience in mechanical engineering based on the courses he taught over thirty-five years, from 1981 in the Department of Mechanical Engineering at the University of Calgary and until his retirement in 2017 from the Department of Mechanical & Materials Engineering at the University of Nebraska, Lincoln. The author and his family sincerely thank many colleagues and friends for their sincere wishes for getting this book published.

The author's family is grateful for the very first encouragement from Professor Neil Popplewell, who is the author's alumus at the Institute of Sound and Vibration Research, University of Souththampton. Professor Popplewell's introduction of his former student was a good start. Many thanks to Professor K. Y. Sze for similar encouragement of getting the book to this stage. Thanks are also due to the author's former students, V. M. Kaladi for allowing the use of figures and tables from his MSc thesis and Jiming Fu, who prepared some drawings for Chapters 2 and 6.

Finally, the author would like to express his gratitude thank the Cambridge Scholars Publishing for their support in the publishing process.

Cho W. S. To

CHAPTER 1

INTRODUCTION

1.1 Background

Vibration or oscillation often refers to the time-varying motion about an equilibrium configuration which is often the stopping configuration of the vibration. Vibrations are important in our daily life. For instance, swings are popularly provided in playgrounds, mechanical clocks and watches keep time by making use of the constant time intervals or periods of the vibrations in their components, vibrations in the musical instruments lead to music, we hear because our eardrums vibrate, etc. Not all vibrations are desirable. Vibrations of land vehicles over bumpy roads and marine vehicles over rough seas may induce motion sickness. Transverse vibrations of high-rise and long-span structures under strong wind can be scary. Excessive vibrations can lead to screw loosening and material fatigue failures which may be catastrophic to the pertinent machines and structures.

1.2 Organization of Presentation

This book is organized into six chapters. The remaining chapters will cover the following content on mechanical vibrations:

- Chapter 2 mentions that the elements capable of holding the kinematic and potential energies are required for mechanical vibrations. The equations of motion for some simple examples of vibration are derived from the energy consideration. Damping which is inevitable in real-world vibrations is also introduced.
- Chapter 3 presents free and forced vibrations of single-degree-of-freedom systems. The equations of motion are derived from not only energy consideration but also Newton's Second Law of Motion. The method for analyzing forced vibrations under non-harmonic and non-periodic excitation forces is introduced.
- Chapter 4 is involved with free and forced vibrations of two-degrees-of-freedom systems. It starts from the forced harmonic vibration and

dynamic absorber. Normal modes and normal mode analysis for forced vibration systems under proportional damping are then introduced. Non-proportional damping is considered by the state vector approach.

- Chapter 5 deals with free and forced vibrations of many degrees of freedom systems. It introduces the Lagrange's equations approach as a generalization of the energy consideration for deriving equations of motion. General treatments of the normal mode analysis and the space vector approach for many degrees of freedom systems are presented.
- Chapter 6 is concerned with continuous systems. Unlike the discrete systems considered in the previous chapters, kinematic and potential energies are distributed in continuous systems which possess an infinite number of degrees of freedom. Free and forced vibration analyses of continuous systems including cables, rods and beams are presented.

CHAPTER 2

OSCILLATORY MOTION

Vibration or oscillation is a continuous exchange of kinematic and potential energy. In Section 2.1 the basic elements capable of storing kinematic and/or potential energy are introduced. The equations of motion for, perhaps, the two simplest vibration systems formed by these elements are derived from energy conservation. Section 2.2 introduces degrees of freedom, discrete systems and continuous systems. Section 2.3 reviews the common forms of damping which dissipate the energy in vibrating systems. In Section 2.4 harmonic and periodic motions are defined. The mathematical procedure for expressing a periodic function into a compound harmonic function is also introduced. Section 2.5 gives a simple example of non-periodic oscillatory motion.

2.1 Elements of Oscillatory Systems

To enable mechanical vibrations, one needs at least a means to store kinetic energy T and a means to store potential energy U. Vibration is a continuous exchange of kinetic energy to and for potential energy. While kinematic energy is held by an object because of its speed, potential energy is held by an object because of its position or deformation.

In mechanical vibrations, the two major forms of potential energy are gravitation energy and elastic energy. From high school physics, we know that when a mass m is raised to a height h above the ground level, the gravitational energy it holds relative to the ground level is mgh where g is the acceleration due to gravity. On the other hand, when a spring with spring constant k is elongated by e relative to its unstretched length, the elastic energy it holds is $ke^2/2$.

In the classical spring-trolley system shown in Figure 2.1(a), a rigid trolley of mass m on light wheels is attached to a rigid wall by a light spring of spring constant k. The upper half of the figure shows the system in which the spring assumes its unstretched length l. The lower half of the figure shows the system in which the trolley is displaced by x towards the right and x would also be the spring extension. Thus, the kinetic and potential energies are

$$T = \frac{1}{2}m\dot{x}^2 \quad \text{and} \quad U = \frac{1}{2}kx^2$$

As usual, a quantity with one and two overdots denote the first and second time derivatives of the quantity under the dot, respectively. In the absence of energy dissipation, the sum of the two energies is a constant, i.e.

$$\frac{1}{2}m\dot{x}^2 + \frac{1}{2}kx^2 = constant \tag{2.1}$$

The time derivative of the above equation is

$$(kx + m\ddot{x})\dot{x} = 0$$

The trivial solution of $\dot{x} = 0$ does not involve any motion. Our concern is

$$\ddot{x} + \frac{k}{m}x = 0 \tag{2.2}$$

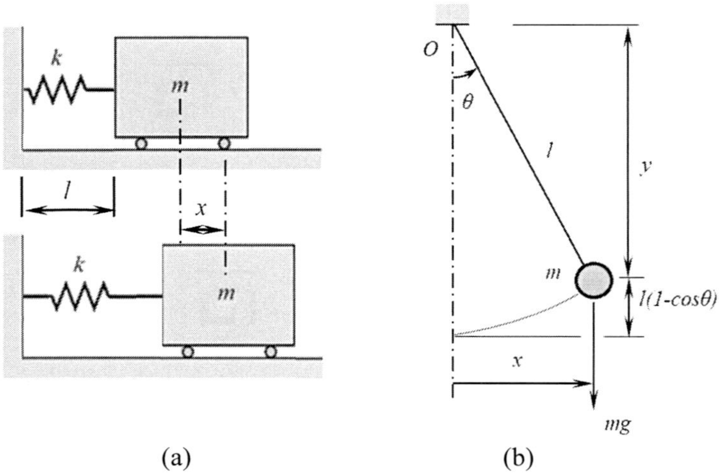

Figure 2.1: (a) A trolley of mass *m* on a smooth horizontal floor, *l* is the unstretched length of the spring. (b) A simple pendulum, *l* is the length of the inextensible string.

The solution to the equation of motion in Equation (2.2) is

$$x = A \sin \omega t + B \cos \omega t \tag{2.3}$$

or, equivalently,

$$x = X \sin(\omega t + \emptyset) \tag{2.4}$$

in which t denotes time, $\omega = \sqrt{k/m}$ is known as the angular velocity of the motion, $X = \sqrt{A^2 + B^2}$ is the amplitude of motion and $\tan \emptyset = B/A$. (A, B) or $(X, \emptyset)$ can be determined by the initial values, i.e. the values at $t = 0$, of x and $\dot{x}$. Equation (2.4) is plotted in Figure 2.2. It can be seen that x oscillates about $x = 0$ and repeats itself at a fixed time interval of $\tau = 2\pi/\omega$ which is known as the period of the vibration. Furthermore, x becomes zero when $\omega t + \emptyset = 0, \pi, 2\pi, \dots$, etc.

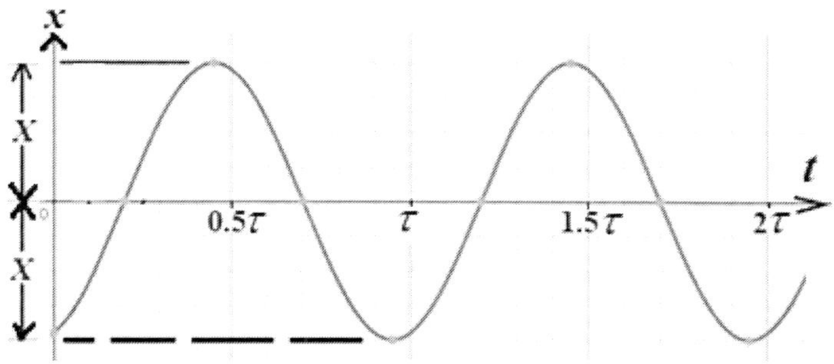

Figure 2.2: A graphical illustration for $X \sin(\omega t + \emptyset)$ where $\tau = 2\pi/\omega$.

Figure 2.1(b) shows the simple pendulum which is another simplest vibration system. A small mass m is tied to a fixed point above it by an inextensible light and taut string of length l. The mass oscillates about the fixed point. When the string makes an angle θ to the vertical, the gravitation energy of the mass relative to its lowest point is

$$U = mgl(1 - cos\theta).$$

On the other hand, the speed of the mass is $l\dot{\theta}$. Thus, the kinematic energy is

$$T = \frac{1}{2} m(l\dot{\theta})^2.$$

Assuming no energy dissipation, the sum of the two energies is a constant, i.e.

$$mgl(1-cos\theta)+\frac{1}{2}m(l\dot{\theta})^2=constant$$

and the time derivative of the energy sum is

$$m(l\dot{\theta})(l\ddot{\theta})+mgl\ sin\theta\ \dot{\theta}=0$$

The trivial static solution given by $\dot{\theta}=0$ can be rejected. By adopting the small-angle approximation $sin\theta\approx\theta$, the above equation reduces to

$$\ddot{\theta}+\frac{g}{l}\theta=0. \qquad (2.5)$$

Equation (2.5) is of the same form as Equation (2.2) and its solution can again be expressed as

$$\theta=\Theta\sin(\omega t+\emptyset) \qquad (2.6)$$

where ω equals $\sqrt{g/l}$ is the angular velocity. Thus, the period of the vibration is $2\pi\sqrt{l/g}$.

2.2 Degrees of Freedom in Mechanical Systems

Degrees of freedom (dofs) are often geometric parameters such as coordinates, displacements and angles of inclination used to describe the time-varying configuration of a vibrating system and, thus, the configurations of all its components. A dof can be dependent or independent. Taking the simple pendulum in Figure 2.1(b) as an example, the horizontal displacement x of the point mass from its lowest point, given by $\theta=0$, can be taken as a dof. However, one can note that $x=l\sin\theta$. If θ is taken as the independent dof, x will be a dependent dof and vice versa. Unless specified otherwise, dofs often refer to independent dofs. Based on the number of dofs it possesses, a vibration system can be classified as a single dof (sdof), a two dof (tdof) or a many dof (mdof) system. While the vibration systems in Figure 2.1 are sdof systems, those in Figure 2.3 are their generalizations and they are tdof systems.

The systems portrayed in Figures 2.1 and 2.3 are also known as discrete vibration systems in the sense that the configuration and, thus, the energies of the system can be fully specified by a finite number of dofs. In these systems, the components carrying the kinematic and elastic energies are typically taken to be rigid and light, respectively. Contrary to the discrete system is the continuous system in which the configuration and, thus, the energies of the system cannot be fully specified by a finite number of dofs.

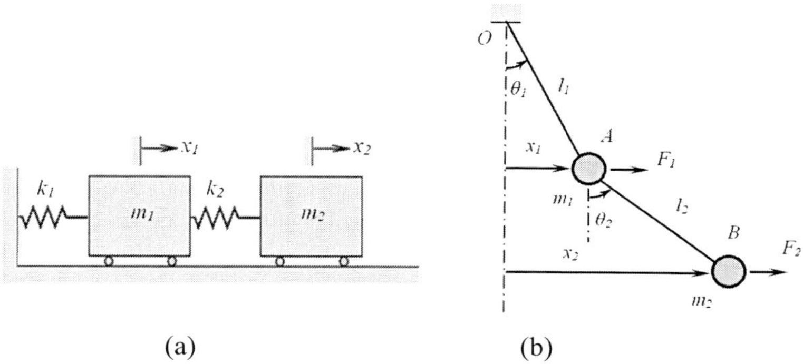

Figure 2.3: (a) A two-trolley system. (b) A double pendulum.

Taking the beam with both ends simply-supported in Figure 2.4 as an example, its kinematic and elastic energies (Rao, 2004, 864) are

$$T = \frac{1}{2}\int \mu \left(\frac{\partial w}{\partial t}\right)^2 dx \quad , \; U = \frac{1}{2}\int EI \left(\frac{\partial^2 w}{\partial x^2}\right)^2 dx$$

in which w, μ and EI are the vertical deflection, the mass per unit length and the flexural rigidity of the beam, respectively. Both energies are distributed in every infinitesimal segment of the beam. In principle, the beam can take up any admissible configurations that satisfy the zero-deflection $w = 0$ and zero-moment $d^2w/dx^2 = 0$ conditions at $x = 0$ and $x = L$ as well as the combinations of these configurations. A set of admissible configurations specified in terms of the w is

$$\{w = \sin(nx/L)\,, n = 1,2,3, \ldots\},$$

see Figure 2.5 for $n = 1$ to 5, and their combination deals to

$$w = \sum_{n=1}^{\infty} c_n \sin(nx/L)$$

in which c_ns are the dofs. Though the summation index can go up to infinity, only the first few terms are of practical importance as the subsequent terms will die out quickly due to energy dissipation or damping to be introduced in Section 2.3. The dof used in a continuous system often refers to the amplitude of a displacement mode. For this reason, these dofs are sometimes termed as modal dofs or coefficients.

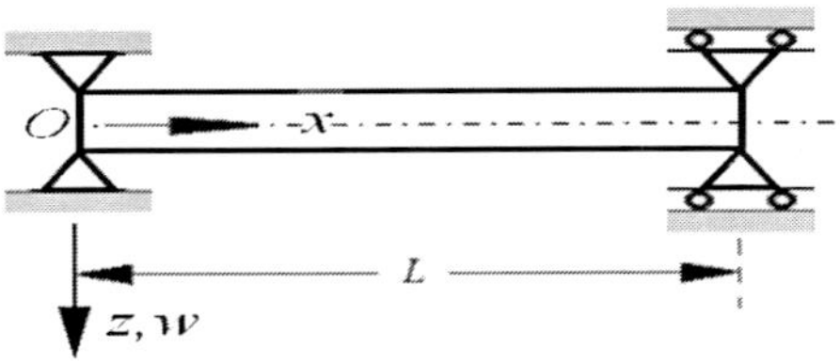

Figure 2.4: A beam with both ends simply supported.

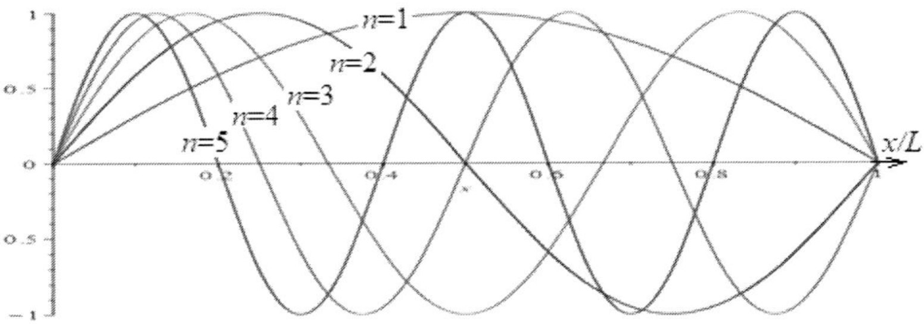

Figure 2.5: Some admissible configurations for the beam with both ends simply supported.

2.3 Viscous, Structural and Coulomb Damping

Damping which opposes the motion and dissipates the energy of a vibration system has not been considered in the discussed examples. In this section, viscous, structural and Coulomb dampings which are the most abundant forms of damping are introduced (Rao, 2004, 139,157) (Thomson and Dahleh, 1993, 27,35,72).

Viscous force or drag is experienced by a body moving in a fluid medium. Consider the viscous fluid between two plates at separation h shown in Figure 2.6. The lower plate is fixed and the upper plate is moving at velocity V. The velocities of the fluid in contact with the two plates will move at the same velocities as the plates. Between the plates, the fluid velocity u can be approximated as linearly varying, i.e.

$$u = \frac{Vy}{h}$$

where y is the distance from the lower fixed plate.

Applying the Newtonian law of viscosity, the horizontal shear stress would be proportional to the velocity gradient du/dy of the fluid, i.e.

$$shear\ stress = \mu \frac{du}{dy} = \frac{\mu V}{h}$$

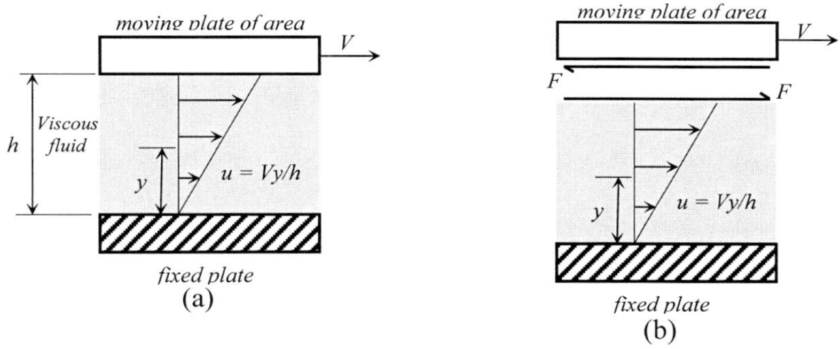

Figure 2.6: (a) Two parallel plates with a viscous fluid in between. (b) The viscous force F experienced by the moving plate.

In the relation, μ is the fluid property known as viscosity and is a measure of how "sticky" the fluid is. Thus, the force resisting the motion of the upper plate of area A is

$$F = A \times shear\ stress = \frac{\mu A}{h} V = cV$$

in which $c = \mu A/h$ is the damping coefficient. In vibration control, viscous damping is mostly realized in a dashpot in which fluid flows through the narrow gap between the piston and the cylinder and/or orifice(s) connecting the two sides of the piston. The velocity of the fluid flow and, thus, the damping force is proportional to the time derivative of the dashpot's extension. A commonly adopted symbol for the dashpot is shown in Figure 2.7 in which the single-dof spring-trolley system is damped by a dashpot of damping coefficient c.

Assuming the displacement takes up the form of Equation (2.4), the viscous damping force would be

$$F = c\dot{x} = cX\omega\, cos(\omega t + \emptyset). \tag{2.7}$$

The energy dissipated by dashpot per vibration cycle is

$$W_d = \int_{one\ cycle} F dx = \int_{t_o}^{t_o+\frac{2\pi}{\omega}} c\dot{x}^2 dt = c(X\omega)^2 \int_{t_o}^{t_o+\frac{2\pi}{\omega}} cos^2(\omega t + \emptyset)\, dt$$
$$= \pi c\omega X^2. \tag{2.8}$$

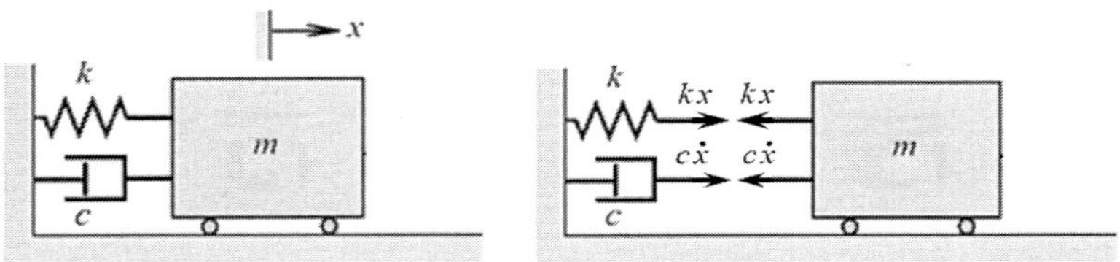

Figure 2.7: A spring-trolley system damped by a dashpot with damping coefficient c.

In structural or material damping, the energy dissipated per vibration cycle takes the following form

$$W_d = \alpha X^2 \tag{2.9}$$

in which α is a constant independent of the angular frequency of the vibration. Figure 2.8 shows a typical hysteresis loop in the stress-strain plot of solid materials. The area inside the loop is the energy dissipated per unit volume of the material in a loading-unloading or vibration cycle. It is trivial that the area inside the loop is proportional to the square of the vibration amplitude if the loop contracts or expands equally in all directions for different vibration amplitudes,

Comparing Equation (2.8) and Equation (2.9) leads to the following equivalent viscous damping coefficient for the structural damping,

$$c_{eq} = \frac{\alpha}{\pi\omega}$$

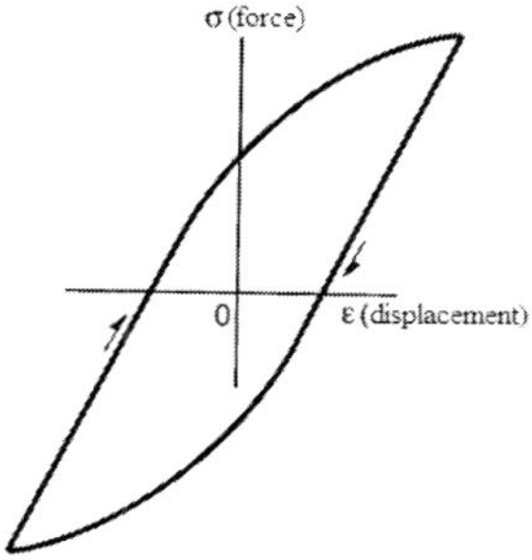

Figure 2.8: Hysteresis effect in the stress-strain and force-displacement plots.

Coulomb damping occurs when two dry solid surfaces rub against each other. The magnitude of the Coulomb damping force or, simply, dry friction is proportional to the normal reaction force between the two contacting surfaces. Assuming the normal reaction force is constant, Figure 2.9 plots the viscous damping force(in solid line) and dry friction force(in chained line) against the velocity. As the forces always oppose the velocity, they are negative when the velocity is positive and vice versa. For the Coulomb damping, the energy dissipation per vibration cycle of amplitude *X* is

$$W_d = 4X \cdot dry\ friction\ force$$

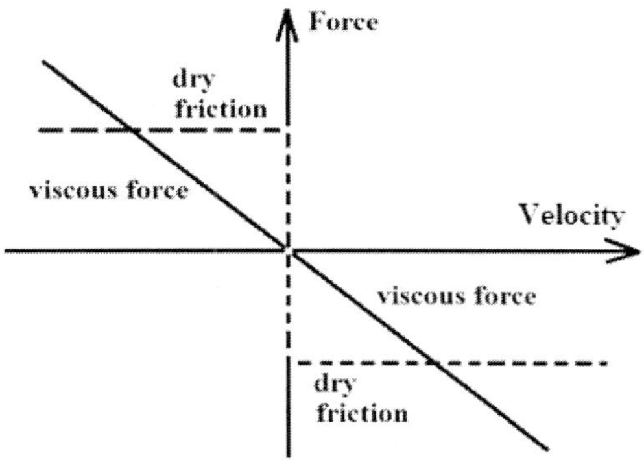

Figure 2.9: Viscous damping force (in solid line) and dry friction force (in chained line) versus the velocity.

2.4 Harmonic and Periodic Motions

An oscillatory motion is termed a harmonic or a simple harmonic motion if the dof, say x, describing the motion can be expressed as

$$x = X\sin(\omega t + \emptyset)$$

Hence, the spring-trolley and the simple pendulum sdof systems discussed in Section 2.1 are undergoing simple harmonic motions. An oscillatory motion x is periodic if it repeats itself at a fixed interval of time, i.e.

$$x(t+\tau) = x(t) \tag{2.10}$$

where τ is the fixed time interval and is known as the period of the motion. It is trivial that harmonic motion is a special type of periodic motion with period $2\pi/\omega$.

A periodic function of period τ can always be expressed as a Fourier series which is a summation series of sine and cosine functions of periods τ/n where $n = 0, 1, 2, 3, \ldots$, i.e.

$$x(t) = \frac{a_0}{2} + \sum_{n=1}^{\infty}\left(a_n\cos(\frac{2n\pi}{\tau}t) + b_n\sin(\frac{2n\pi}{\tau}t)\right) \tag{2.11}$$

and the first term arises from $n = 0$ [1,2]. In other words, a periodic motion can be regarded as a compound harmonic motion. Furthermore,

$$a_n\cos(\frac{2n\pi}{\tau}t) \quad \text{and} \quad b_n\sin(\frac{2n\pi}{\tau}t)$$

are termed as the n-th harmonic components of $x(t)$. To find a_ns and b_ns for a given $x(t)$, both sides of (2.11) can be multiplied with 1, $\cos(\frac{2m\pi}{\tau}t)$ and $\sin(\frac{2m\pi}{\tau}t)$, and integrated from t_o to $t_o+\tau$, namely

$$\begin{aligned}
\int_{t_o}^{t_o+\tau} x(t)dt &= \int_{t_o}^{t_o+\tau} a_0 dt \\
&\quad + \sum_{n=1}^{\infty}\int_{t_o}^{t_o+\tau}\left(a_n\cos(\frac{2n\pi}{\tau}t) + b_n\sin(\frac{2n\pi}{\tau}t)\right)dt, \\
\int_{t_o}^{t_o+\tau} x(t)\cos(\frac{2m\pi}{\tau}t)\,dt &= \int_{t_o}^{t_o+\tau} a_0\cos(\frac{2m\pi}{\tau}t)\,dt \\
&\quad + \sum_{n=1}^{\infty}\int_{t_o}^{t_o+\tau}\cos(\frac{2m\pi}{\tau}t)\left(a_n\cos(\frac{2n\pi}{\tau}t) + b_n\sin(\frac{2n\pi}{\tau}t)\right)dt, \\
\int_{t_o}^{t_o+\tau} x(t)\sin(\frac{2m\pi}{\tau}t)\,dt &= \int_{t_o}^{t_o+\tau} a_0\sin(\frac{2m\pi}{\tau}t)\,dt \\
&\quad + \sum_{n=1}^{\infty}\int_{t_o}^{t_o+\tau}\sin(\frac{2m\pi}{\tau}t)\left(a_n\cos(\frac{2n\pi}{\tau}t) + b_n\sin(\frac{2n\pi}{\tau}t)\right)dt
\end{aligned} \tag{2.12}$$

in which m is a non-zero positive integer and t_o is an arbitrary instant of time. By invoking

$$\int_{t_o}^{t_o+\tau} \cos(\frac{2m\pi}{\tau} t)\, dt = \int_{t_o}^{t_o+\tau} \sin(\frac{2n\pi}{\tau} t)\, dt = \int_{t_o}^{t_o+\tau} \cos(\frac{2m\pi}{\tau} t) \sin(\frac{2n\pi}{\tau} t)\, dt = 0\,,$$

$$\int_{t_o}^{t_o+\tau} \cos(\frac{2m\pi}{\tau} t) \cos(\frac{2n\pi}{\tau} t)\, dt = \int_{t_o}^{t_o+\tau} \sin(\frac{2m\pi}{\tau} t) \sin(\frac{2n\pi}{\tau} t)\, dt = \frac{\delta_{mn}}{2}$$

where δ_{mn} is equal to unity when $m = n$ and zero otherwise, the three relations in Equation (2.12) yield

$$a_0 = \frac{2}{\tau} \int_{t_o}^{t_o+\tau} x(t) dt \;,\; a_n = \frac{2}{\tau} \int_{t_o}^{t_o+\tau} x(t) \cdot \cos(\frac{2n\pi}{\tau} t)\, dt,$$

$$b_n = \frac{2}{\tau} \int_{t_o}^{t_o+\tau} x(t) \cdot \sin(\frac{2n\pi}{\tau} t)\, dt,$$

respectively.

2.5 Non-Periodic Motion

Contrary to the periodic motion, the non-periodic motion is a type of motion that does not repeat itself as described in Equation (2.10).

A typical example of non-periodic motion is a ball dropping vertically downward from a height. The ball will be rebounded by the floor. Let the coefficient of restitution be $e < 1$, the velocity (upward as positive) of the ball before the first rebound be $-u$ and the air resistance be negligible, the following list shows the velocity, the time of flight and the maximum height reached by the ball after each rebound.

	velocity after the rebound	time of flight after the rebound	maximum height after the rebound
Rebound 1	eu	$2eu/g$	$(eu)^2/(2g)$
Rebound 2	e^2u	$2e^2u/g$	$(e^2u)^2/(2g)$
:	:	:	:
Rebound n	e^nu	$2e^nu/g$	$(e^nu)^2/(2g)$

The motion of the ball is non-periodic because both the time of flight and the maximum height of the ball after each rebound are diminishing. Of course, the motion becomes periodic if $e = 1$ which implies no energy is lost in the rebound.

References

Rao, S.S. 2004. Mechanical Vibrations, Fourth Edition. New Jersey: Pearson.

Thomson, W. T., and Dahleh, M.D. 1993. Theory of Vibration with Applications, Fifth Edition. Englewood Cliffs, New Jersey: Prentice-Hall.

Exercise Questions

Q1. A light rigid rod OA of length $3l$ is hanged to a smooth pivot O. A mass m is attached to A whilst two horizontal springs of spring constant k are attached to the rod at l and $2l$ from O as shown in Figure 2Q1. When the rod is vertical, the spring force is zero. Assuming that the swinging angle is small, derive the kinematic, gravitational and elastic energies of the system. Hence, find the frequency of the oscillation.

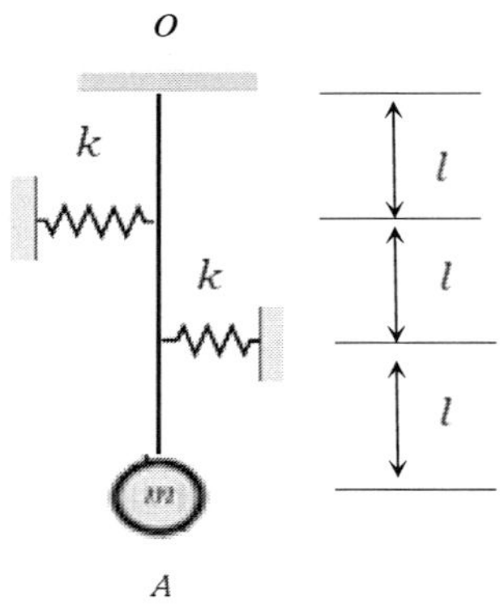

Figure 2Q1: A light rod OA pivoted at O with attached mass and springs.

Q2. A pendulum is formed by hanging a uniform rigid rod of mass m and length l to a smooth pivot. Assuming that the swinging angle is small, derive the kinematic and potential energies of the rod. Hence, find the frequency of the oscillation.

Q3. The initial displacement and initial velocity of a simple harmonic motion of frequency 5 Hz are 0.01 m and 0.1 m/s, respectively. Determine the displacement solution.

Q4. The following figure shows a rectangular waveform $x(t)$ of period $2\pi/\omega$ and magnitude A. Determine the Fourier series of $x(t)$.

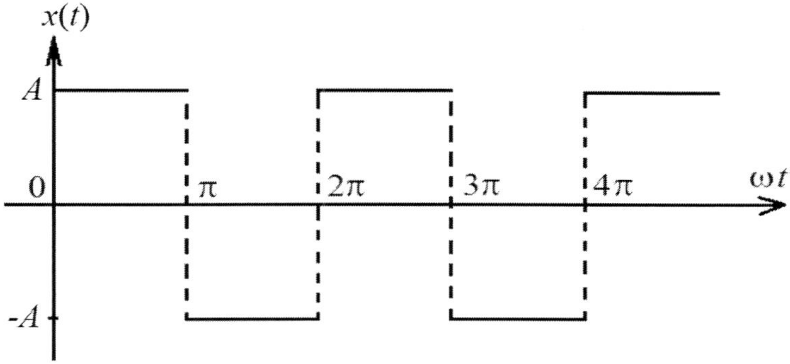

Figure 2Q4: A rectangular waveform $x(t)$.

CHAPTER 3

FREE AND FORCED VIBRATION OF SINGLE DEGREE OF FREEDOM SYSTEMS

In translating the oscillatory or oscillatory mechanical systems in the physical world into mathematical models in the conceptual world there is always a question of how accurate and realistic of representation of the physical system. It is generally logical to approach the problem by choosing a simplest model and if one desires, more realistic model can be formulated at a later stage. The simplest model is generally governed by the economy of solution. This simplest model can often provide information and insight to the problem at hand.

The present chapter therefore begins with the presentation of translational and rotational elements of oscillatory systems, formulation and analysis of simplest dynamic systems. The single degree-of-freedom (dof) systems without applied forces are dealt with in Section 3.1. Section 3.2 deals with single dof linear systems under harmonic forces. Section 3.3 introduces single dof linear systems under periodic forces. Section 3.4 is concerned with non-harmonically and non-periodic forced single dof linear systems. Questions and their solutions are included in Section 3.5.

3.1 Free Vibration of Single Degree of Freedom Systems

The presentation in this section is as follows. First, the systems in the physical world as shown in Figure 3.1 are given. Second, their corresponding models in the conceptual world are presented in Figure 3.2. Third, the formulation and analysis of one of the models in the conceptual world is selected to illustrate the solution steps.

Stage 1: ***Systems in Physical World***

There are many systems in the engineering physical world. It suffices to give two common examples for the purpose of illustrating the process of modelling. These two systems in the physical world are the automobile and television tower as sketched in Figure 3.1.

Stage 2: ***Systems in Conceptual World***
The systems in Figure 3.1 may be represented by the simplest mathematical models, known as *lumped-parameter models* or *discrete systems*, in the conceptual world. They are illustrated in Figure 3.2.

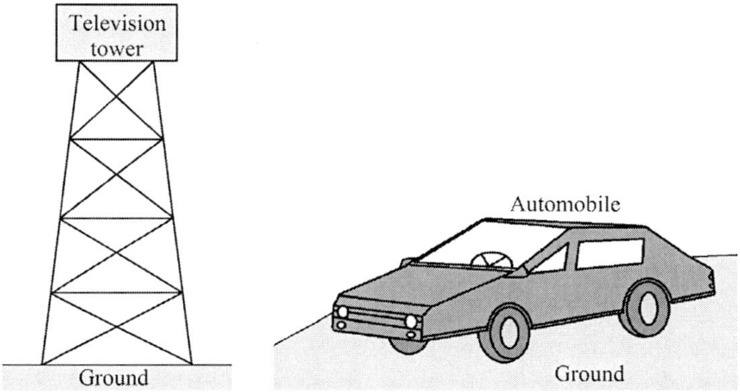

Figure 3.1: Systems in physical world.

Stage 3: ***Formulation and Solution***
The equation of motion governed each of the conceptual world models sketched in Figure 3.2 may now be derived. In general, the equation of free vibration for a single dof system may be obtained by one of the following approaches:

- Definition of simple harmonic motion,
- Knowledge of oscillatory motion,
- Law of conservation of energy,
- Newton's second law of motion, and
- Method of virtual work or virtual power.

Consider the discrete system on the left-hand side (lhs) in Figure 3.2. The latter and its free-body diagram (FBD) are included in Figure 3.3.

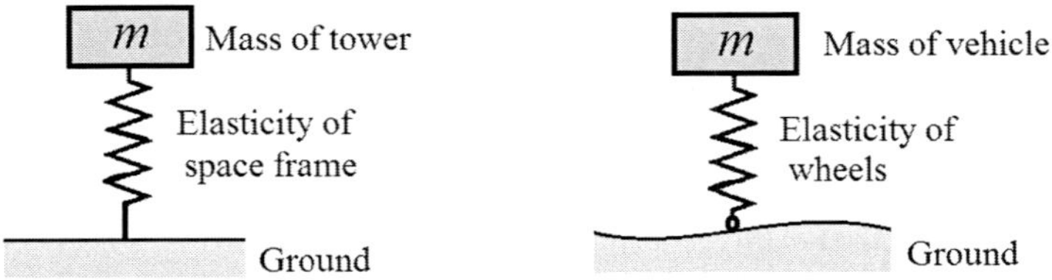

Figure 3.2: Single dof models of systems in Figure 3.1.

By Newton's second law of motion for the lumped mass,

$$m\ddot{x} = \sum F = mg - k(\Delta + x) - c\dot{x},$$

in which *m* is the mass, *k* spring constant, and *c* damping coefficient of the system, respectively. The symbol $\sum F$ denotes summing of the forces. The spring force $k\Delta$ is equal to the gravitational force *mg* acting on the mass *m*. That is, $mg = k\Delta$.

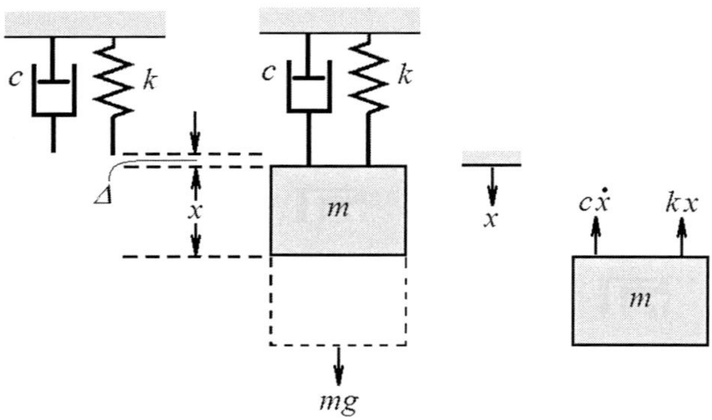

Figure 3.3: Single dof system and free-body diagram.

Therefore, the equation of motion for the above system is

$$m\ddot{x} + c\dot{x} + kx = 0. \tag{3.1}$$

The first, second and third terms on the lhs of Equation (3.1) are the inertia, damping, and restoring forces, respectively.

One can solve Equation (3.1) by the Laplace transform method but in the following another independent approach is employed to provide one more alternative.

Let the solution of the above system be $x = e^{st}$, where s is a constant parameter. The basis for such an assumption is that physically one is dealing with oscillatory motion and mathematically the governing equation of motion is a second order linear differential equation.

Performing the differentiations of x with respect to time t and substituting the displacement, velocity, and acceleration terms into Equation (3.1), one obtains

$$(ms^2 + cs + k)e^{st} = 0$$

which is satisfied for all values of t (physically, it implies that the system undergoes vibrating motion) when

$$s^2 + \frac{c}{m}s + \frac{k}{m} = 0 \qquad \text{or}$$

$$s^2 + 2\zeta\omega_n s + \omega_n^2 = 0. \tag{3.2}$$

where $c/m = 2\zeta\omega_n$ and $k/m = \omega_n^2$. In the foregoing, ζ is known as the *damping ratio* and ω_n is the *undamped natural frequency* or simply referred to as *natural frequency* of the system. Equation (3.2) is known as the *characteristic equation* (c.e.) and has two roots since it is a quadratic equation. The two roots are

$$s_1, s_2 = -\zeta\omega_n \pm i\omega_n\sqrt{1-\zeta^2}. \tag{3.3}$$

- For *lightly damped* system, that is $0 < \zeta < 1$,

$$s_1, s_2 = -\alpha \forall i\omega_d \tag{3.4}$$

where $\alpha = \zeta\omega_n$ and $\omega_d = \omega_n\sqrt{1-\zeta^2}$ is known as the *damped natural frequency* or *resonant frequency* of the system.

The general solution to Equation (3.1) can be written as

$$x = A_1 e^{(-\alpha + i\omega_d)t} + A_2 e^{(-\alpha - i\omega_d)t}. \tag{3.5}$$

or factoring the exponential term such that it becomes

$$x = e^{-\alpha t}(A_1 e^{i\omega_d t} + A_2 e^{-i\omega_d t})$$

Where $e^{i\omega_d t} = cos\omega_d t + i\ sin\omega_d t,\quad e^{-i\omega_d t} = cos\omega_d t - i\ sin\omega_d t$
and then it can be rewritten as

$$x = e^{-\alpha t}[(A_1 + A_2)\ cos\omega_d t + i\ (A_1 - A_2) sin\omega_d t]. \tag{3.6}$$

Introducing the following notation,

$$A_1 + A_2 = A \cos\phi\ ,\ \ i\ (A_1 - A_2) = A\ sin\ \phi$$

such that Equation (3.6) becomes

$$x = Ae^{-\alpha t}(\cos\phi\ cos\omega_d t + \ sin\ \phi sin\omega_d t).$$

But by using the trigonometric identity,

$$\cos(\phi - \omega_d t) = cos(\omega_d t - \phi) = cos\omega_d t + \ sin\ \phi sin\omega_d t,$$

the solution can be simplified to

$$x = Ae^{-\alpha t}cos(\omega_d t - \phi) \tag{3.7}$$

where A and ϕ are constants of integration and are dependent of the initial displacement and initial velocity because A and ϕ are defined in terms of A_1 and A_2 in the foregoing.

Remark 3.2.1
In the special case in which *damping is critical*, that is, $\zeta = 1$, the solution to Equation (3.1) can be shown to be

$$x = (A_1 + tA_2)e^{-\alpha t}.$$

Remark 3.2.2
For the case in which $\zeta > 1$, that is *over-damped or heavily damped* free vibration, Equation (3.3) gives two real roots and the solution becomes

$$x = e^{-\alpha t}\left(A_1 e^{\omega_n\sqrt{\zeta^2 - 1}\,t} + A_2 e^{-\omega_n\sqrt{\zeta^2 - 1}\,t}\right).$$

Remark 3.2.3
In general, the lightly damped free vibration is the most common type of free vibration that can occur in many mechanical dynamic systems.

3.2 Single Degree-of-Freedom Systems under Harmonic Forces

Having derived the equation of free vibration for a single dof oscillatory system, it is logical to study the vibration of a forced single dof system. In practice, machines and structural systems are often accompanied by various dynamic forces. One type of these dynamic forces is the harmonic force.

- Common sources of harmonic forces in single dof systems are:
  (a) reciprocating machines,
  (b) unbalance in rotating machines, and
  (c) support motions.
- One of the main objectives of the study is to obtain the amplitude of vibration developed in the system.
- To prevent machine failure or catastrophic consequence, *resonance* (that is, when the applied frequency is equal to the natural frequency of the system) in the system is to be avoided.

3.2.1 Reciprocating machines

Consider the single dof system, shown in Figure 3.4, which represents a simplified reciprocating machine. The equation of motion of such a machine may be derived as

$$m\ddot{x} + c\dot{x} + kx = f(t) = F\sin\omega t\,, \tag{3.8}$$

where F is the amplitude of the harmonic force $f(t)$ or for simplicity, written as f.

- The solution to this second order differential equation consists of two parts, the *complementary function or transient response*, which is the solution of the homogeneous equation, and the *particular integral*.

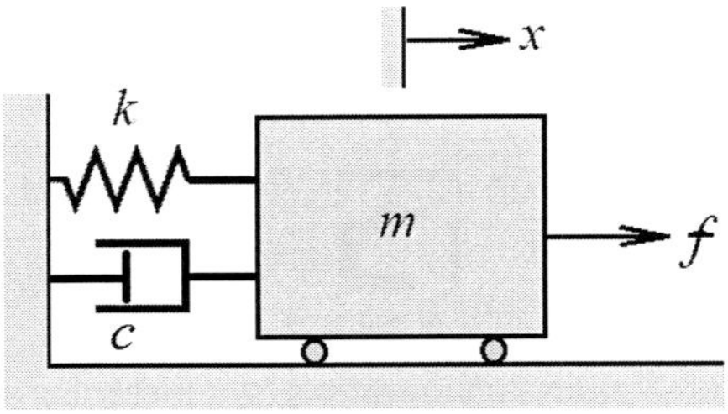

Figure 3.4: Harmonically forced single dof system.

The transient response has been discussed in Section 3.1. The present objective is to obtain the particular integral. Since it is assumed that the system has positive damping and therefore the response occurs at a later time or the response is *delayed* by a phase angle φ with respect to the applied force *f*, so that the solution or *steady state displacement* may be written as

$$x = X\sin(\omega t - \varphi) \tag{3.9}$$

where X is the amplitude of the displacement and not to be confused with the Laplace transform of the instantaneous displacement x.

At this stage, it may be appropriate to note that since linear single dof systems are being considered presently the response of the system to the applied force is proportional to the type of excitation. This is why when the applied force is of the sinusoidal form the response is also in that form, except that it is delayed by the phase angle φ since the system has positive damping.

The vector relationship for forced vibration with viscous damping is shown in Figure 3.5 below. With reference to the vector diagram, one has

$$F^2 = (kX - m\omega^2 X)^2 + (c\omega X)^2$$

and therefore,

$$X = \frac{F}{\sqrt{(k - m\omega^2)^2 + (c\omega)^2}}, \qquad \tan\varphi = \frac{c\omega}{k - m\omega^2}. \quad (3.10a, b)$$

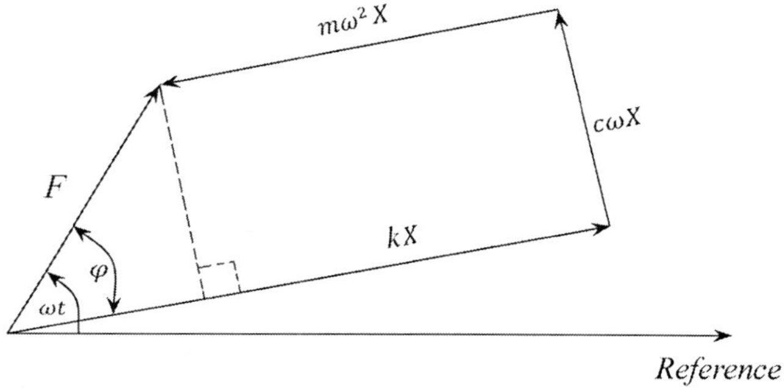

Figure 3.5: Vector representations of forces in forced single dof system.

Now that the amplitude X and phase φ are determined Equation (3.9) can be assembled. For the lightly damped case, the complete solution to Equation (3.8) is

$$x = e^{-\alpha t}(A_1 e^{i\omega_d t} + A_2 e^{-i\omega_d t}) + X\sin(\omega t - \varphi) \tag{3.11}$$

in which the two constants A_1 and A_2 are found by applying the initial conditions.

It may be mentioned that if damping is disregarded in the system, the complete solution reduces to

$$x = (A_1 e^{i\omega_n t} + A_2 e^{-i\omega_n t}) + X\sin(\omega t). \tag{3.12}$$

where now $X = \frac{F}{k - m\omega^2}$ and $\varphi = 0$.

- In this case Equation (3.12) assumes the applied frequency ω is not equal to the natural frequency ω_n of the system. If the applied frequency is equal to the natural frequency such that the equation of motion becomes

$$\ddot{x} + \omega_n^2 x = \frac{F}{m}\sin\omega_n t. \tag{3.13}$$

The complete solution to Equation (3.13) is

$$x = (A_1 e^{i\omega_n t} + A_2 e^{-i\omega_n t}) - \frac{F}{2m\omega_n} t\cos(\omega_n t). \tag{3.14}$$

Note that the response is growing with time in Equation (3.14).

Now, the dimensionless form of Equation (3.10) can be obtained dividing the numerator and denominator on the rhs by k as

$$X = \frac{F/k}{\sqrt{[1-(m\omega^2/k)]^2 + (c\omega/k)^2}}, \qquad \tan\varphi = \frac{c\omega/\mathrm{k}}{1-(m\omega^2/k)}$$

so that upon moving the resulting numerator on the rhs to the denominator on the lhs and making use of the definitions $c/m = 2\zeta\omega_n$ and $k/m = \omega_n^2$, one has

$$\frac{Xk}{F} = \frac{1}{\sqrt{[1-(\omega/\omega_n)^2]^2 + [2\zeta(\omega/\omega_n)]^2}}, \qquad \tan\varphi = \frac{2\zeta(\omega/\omega_n)}{1-(\omega/\omega_n)^2}.$$

Writing the ratio of frequencies $r = \omega/\omega_n$, one obtains

$$\frac{Xk}{F} = \frac{1}{\sqrt{(1-r^2)^2 + (2\zeta r)^2}}, \qquad \tan\varphi = \frac{2\zeta r}{1-r^2}. \qquad (3.15a, b)$$

The *magnification factor* and *phase angle* given by Equations (3.15*a, b*) are included in Figure 3.6 for three different damping ratios.

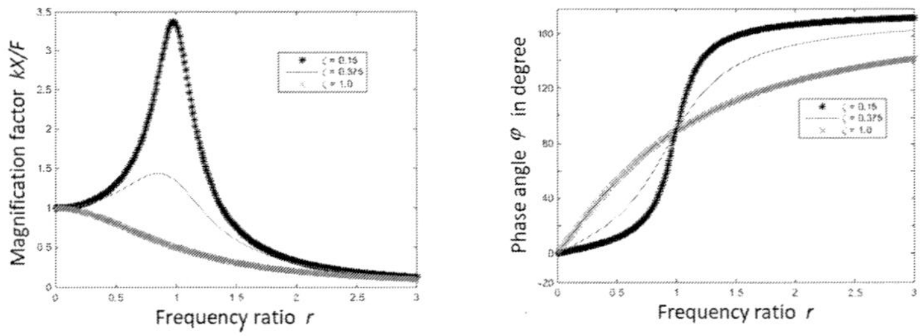

Figure 3.6: Response and phase relationships of single dof system.

- Note that in practice in order to avoid resonance or significantly reduce the vibration of the system the applied frequency should be 2 or more times of the natural frequency, as indicated in Figure 3.6.

3.2.2 Unbalance in rotating machines

Unbalance in rotating machines is common. A schematic diagram for a rotating unbalance is shown in Figure 3.7.

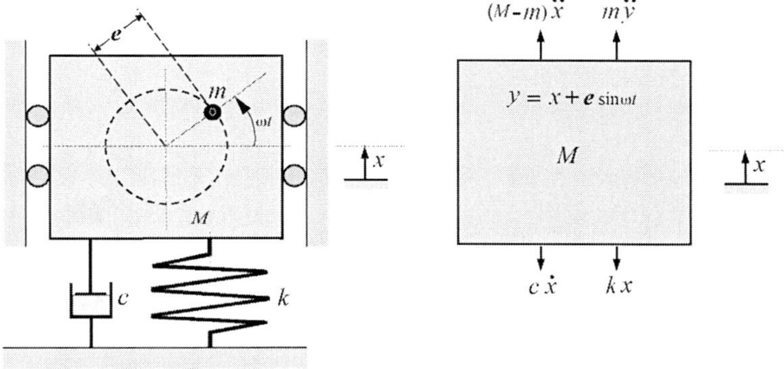

Figure 3.7: Rotating unbalance in a machine.

- The rotating unbalance is represented by the eccentric mass m with eccentricity e which is rotating with an angular velocity ω
- Let x is the *displacement of the non-rotating* mass $M - m$ from the *equilibrium* position.
- The displacement of the rotating mass m is $x + e \sin \omega t$.

By making use of the FBD in Figure 3.7, one has

$$(M - m)\ddot{x} + m \frac{d^2(x + \sin \omega t)}{dt^2} = -c\dot{x} - kx$$

upon operating the differentiation and rearranging, one obtains

$$M\ddot{x} + c\dot{x} + kx = (me\omega^2) \sin \omega t. \quad (3.16)$$

This equation is similar in form to Equation (3.8), except F in the latter equation is replaced by $me\omega^2$. This means that the amplitude and phase of vibration in Equation (3.16) can be obtained as Equation (3.10) to give

$$X = \frac{me\omega^2}{\sqrt{(k - m\omega^2)^2 + (c\omega)^2}}, \qquad \tan\varphi = \frac{c\omega}{k - m\omega^2}. \quad (3.17a, b)$$

The dimensionless forms of the amplitude and phase of the vibration can therefore be similarly obtained as Equations (3.15*a*, *b*) such that

$$\frac{XM}{me} = \frac{r^2}{\sqrt{(1 - r^2)^2 + (2\zeta r)^2}}, \qquad \tan\varphi = \frac{2\zeta r}{1 - r^2}. \quad (3.18a, b)$$

The dimensionless ratio in Equation (3.18*a*) is included in Figure 3.8 for direct observation while the phase angle in Equation (3.18*b*) is identical to that in Equation (3.15*b*) and therefore it has already been presented in Figure 3.6.

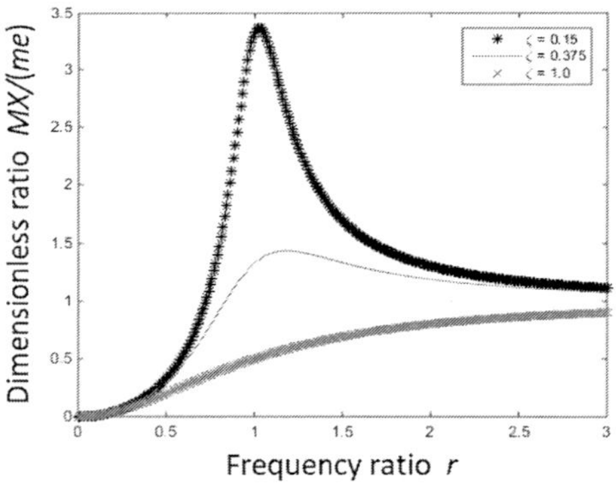

Figure 3.8: Response and phase relationships of unbalance system.

- With reference to Figure 3.8, in order to avoid resonance or significantly reduce the dimensionless ratio (of total mass and amplitude of displacement to unbalance) one should keep the latter to within 0.5 the frequency ratio (the frequency of the rotating mass to the natural frequency of the total mass).

3.2.3 Response of systems under support motion

In *packaging* (secondary equipment in a primary system), sensors attached to vibrating machines or the wing of an airplane which is represented as sinusoidal excitation due to oscillatory motion of the wing, and building as a single spring-mass system under earthquake excitation which may be simplified as a sinusoidal motion, are examples in this category of single dof systems.

The schematic diagram of such a single dof system is shown in Figure 3. 9 in which the FBD is also included. By using the FBD, one has

$$m\ddot{x} = -\,(x-y) - (x-y) = 0.$$

Upon rearranging and let the relative displacement or drift $z = x - y$, one obtains

$$m\ddot{z} + c\dot{z} + kz = -m\ddot{y} = (m\omega^2 Y)\sin\omega t. \qquad (3.19)$$

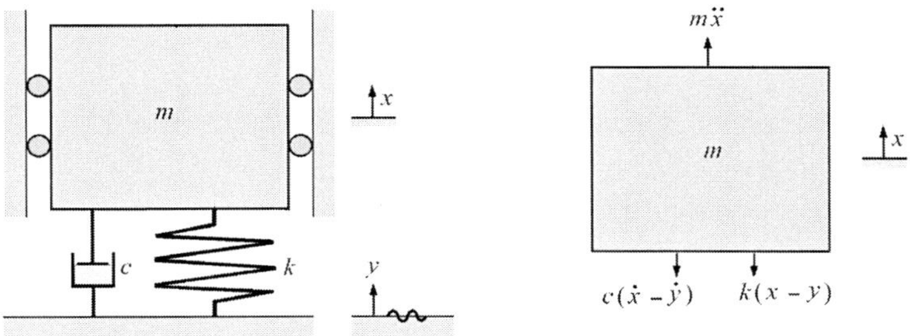

Figure 3.9: Single dof system under support motion.

Equation (3.19) is similar in form to Equation (3.16). Therefore, similar to Equation (3.18) one can obtain

$$\frac{Z}{Y} = \frac{r^2}{\sqrt{(1-r^2)^2 + (2\zeta r)^2}}, \qquad \tan\varphi = \frac{2\zeta r}{1-r^2}. \qquad (3.20a, b)$$

in which now $\omega_n = \sqrt{k/m}$ instead of $\omega_n = \sqrt{k/M}$. The graphical representation of Equation (3.20*a*) is similar to Figure 3.8 except that now the ordinate is Z/Y. It is appropriate to note that the phase angle defined in Equation (3.20*b*) is identical to those in Equations (3.15b) and (3.18b).

3.3 Single Degree-of-Freedom Systems under Periodic Forces

The harmonic excitation considered in the last section is a special case of the periodic forces. Mathematically, any non-harmonic forces can be expressed as a *Fourier series*.

Consider the single dof system in Figure 3.4 in which the forcing function is now in the form of periodic square waves as shown in Figure 3.10 in which the period is 2π, that is $f(t+2\pi)=f(t)$.

The equation of motion is similar to Equation (3.8), except that the rhs is replaced by the periodic square wave excitation

$$f(t)=\frac{4F}{\pi}\left[\sin t+\frac{1}{3}\sin(3t)+\frac{1}{5}\sin(5t)+\ \dots\right]. \qquad (3.21)$$

Before the steady-state response of the sdof system under the periodic square wave excitation is determined, the following will show the periodic square waves in Figure 3.10 becomes Equation (3.21).

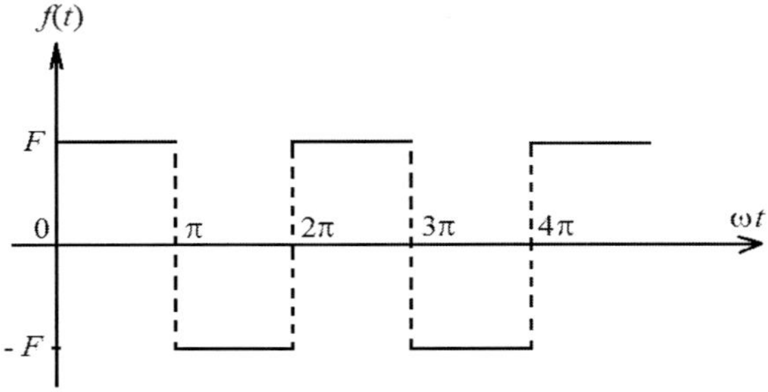

Figure 3.10: Force in form of periodic square waves.

With reference to Figure 3.10, it is clear that *f*(*t*) is an odd function and therefore the coefficient a_n in the Fourier series,

$$f(t)=\frac{a_o}{2}+\sum_{n=1}^{\infty}[\,a_n\cos{(nt)}+b_n\sin(nt)]$$

is $a_n = 0$ with $n = 0, 1, 2, \dots, \infty$ and the coefficient

$$b_n = \frac{2}{2\pi}\int_0^{2\pi} f(t)\sin(nt)\,dt\,,$$

in which the period of square waves is 2π and $\omega = 1$. Thus,

$$b_n = \frac{1}{\pi}\int_0^{\pi} F\sin(nt)\,dt + \frac{1}{\pi}\int_{\pi}^{2\pi}(-F)\sin(nt)\,dt$$

$$b_n = \frac{F}{\pi}\left[-\frac{\cos(nt)}{n}\right]_0^{\pi} + \frac{(-F)}{\pi}\left[-\frac{\cos(nt)}{n}\right]_{\pi}^{2\pi}$$

$$= \frac{F}{n\pi}[-\cos(n\pi) + \cos(n\,.0)] + \frac{(-F)}{n\pi}[-\cos(2n\pi) + \cos(n\pi)]$$

$$= \frac{F}{n\pi}[-\cos(n\pi) + 1] + \frac{(-F)}{n\pi}[-1 + 2\sin^2(n\pi) + \cos(n\pi)]$$

since $\cos(2n\pi) = 1 - 2\sin^2(n\pi)$. But $\sin(n\pi) = 0$. Therefore,

$$b_n = \frac{2F}{n\pi}[-\cos(n\pi) + 1\,] = \begin{cases} 0 & \text{for } n \text{ even} \\ \dfrac{4F}{n\pi} & \text{for } n \text{ odd}. \end{cases}$$

Therefore,

$$f(t) = \frac{4F}{\pi}\left[\sin t + \frac{1}{3}\sin(3t) + \frac{1}{5}\sin(5t) + \dots\right].$$

This is Equation (3.21).

By making use of Equations (3.9) and (3.10), the steady-state displacement becomes

$$x(t) = \frac{4F}{\pi}\left\{\sum_{s=1}^{\infty}\frac{\sin\left[(2s-1)t - \phi_{2s-1}\right]}{(2s-1)\sqrt{[k - m(2s-1)^2]^2 + [c(2s-1)]^2}}\right\},\tag{3.22a}$$

In which

$$\tan\phi_{2s-1} = \frac{c(2s-1)}{k - m(2s-1)^2},\tag{3.22b}$$

where *m, k*, and *c* are the mass, spring constant, and damping coefficient of the sdof system, respectively.

3.4 Single Degree-of-Freedom Systems under Non-Harmonic and Non-Periodic Forces

The impulse excitation and response, and an arbitrary excitation and response of a single dof system are presented. In addition, response of a single dof system under initial conditions is included. As pointed out in Chapter 2, all functions considered in this book are deterministic in the sense that no statistical analysis is required. However, it suffices to mention that for some solutions of response of single dof systems to the excitations that are difficult to determine analytically, numerical methods are often employed.

3.4.1 Impulse excitation and response

Starting from Newton's second law of motion

$$\vec{F} = m\vec{a} = \frac{d\,(m\vec{v})}{dt}$$

where $\vec{F}$ is the force, m the mass, $\vec{a}$ the acceleration, $\vec{v}$ the velocity, and $m\vec{v}$ the linear momentum. One can multiply both sides of this equation by dt and integrating between times t_1 and t_2 such that

$$\int_{t_1}^{t_2} \vec{F} dt = \boldsymbol{I}_{1\to 2} = \int_{\vec{v}_1}^{\vec{v}_2} d\,(m\vec{v}) = m\vec{v}_2 - m\vec{v}_1 \tag{3.23}$$

in which $\vec{v}_1$ and $\vec{v}_2$ are the velocities at times t_1 and t_2, respectively. The integral on the lhs is known as the *impulse in the time interval between* t_1 *and* t_2. If the impulse $\boldsymbol{I}_{1\to 2}$ is equal to unity and it is denoted by $\overline{\boldsymbol{I}}_{1\to 2}$ which is called the *unit impulse.*

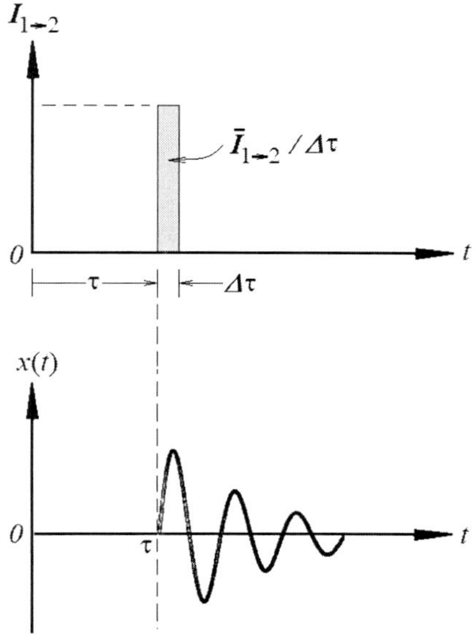

Figure 3.11: Impulse and impulse response function.

Now, consider the response of a lightly damped single dof system to a unit impulse excitation at $t = 0$. Starting with the free vibration as in Section 3.1 and the initial conditions $x(0) = 0,\ \dot{x}(0) = \bar{\boldsymbol{I}}_{1\to2}/m$ which is obtained by making use of Equation (3.19), so that the response of a lightly damped single dof system to a unit impulse $\bar{\boldsymbol{I}}_{1\to2}$ becomes

$$x = \frac{\bar{\boldsymbol{I}}_{1\to2}\, e^{-\zeta\omega_n t}}{m\omega_d} \sin\omega_d t = \bar{\boldsymbol{I}}_{1\to2} h(t), \quad \omega_d = \omega_n\sqrt{1-\zeta^2}, \quad (3.24)$$

in which $h(t) = \dfrac{e^{-\zeta\omega_n t}}{m\omega_d} \sin\omega_d t$ is called the *impulse response function.*

In Figure 3.11, the impulse of magnitude $\bar{\boldsymbol{I}}_{1\to2}/\Delta\tau$ with a duration $\Delta\tau$ becomes infinite in the limit when $\Delta\tau \to 0$. On the other hand, $\bar{\boldsymbol{I}}_{1\to2}$ is considered finite. This leads to the concept of the *delta function* . Mathematically, a delta function at $t = \tau$ is denoted by the symbol $\delta(t - \tau)$ and has the following properties:

$$\delta(t-\tau) = 0\ , \qquad \text{for all } t \neq \tau,\ 0 < \tau < \infty$$

$$\int_0^{\infty} \delta(t-\tau)dt = 1.$$

3.4.2 Arbitrary excitation and response

With the response to a unit impulse established in Equation (3.24) it is possible to derive the equation of the response of the lightly damped single dof system to an arbitrary excitation $f(t)$ which is shown in Figure 3.12.

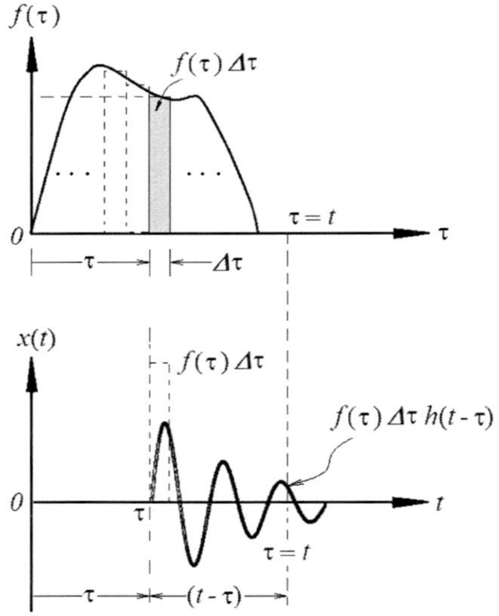

Figure 3.12: Arbitrary excitation as a series of impulses and response.

Suppose one of the impulses at $t = \tau$ has its strength as

$$\boldsymbol{I}_{1\to 2} = f(\tau)\Delta\tau$$

and its contribution to the response at time t is governed by the elapsed time $(t-\tau)$ such that

$$x(t) = f(\tau)\Delta\tau\, h(t-\tau)$$

where $h(t-\tau)$ is the response to a unit impulse? The total response at time t is determined by summing all the responses due to the impulses at all time τ. That is,

$$x(t) = \sum f(\tau)\Delta\tau\, h(t-\tau).$$

In the limit, $\Delta\tau \rightarrow dt$ so that the summation is replaced by the integration, one has

$$x(t) = \int_0^t f(\tau)\, h(t-\tau)d\tau\,. \qquad (3.25)$$

Equation (3.25) is known as the *convolution integral* or *superposition integral* or *Duhamel integral*.

3.4.3 Initial conditions and response

For completeness, the case of spring-mass dropping through a vertical distance of *d*, as shown in Figure 3.13 is considered here. This is a simple case of non-harmonically forced vibration. Use of such a model can be found in analysis of the free fall of a landing helicopter or aircraft and packaging design.

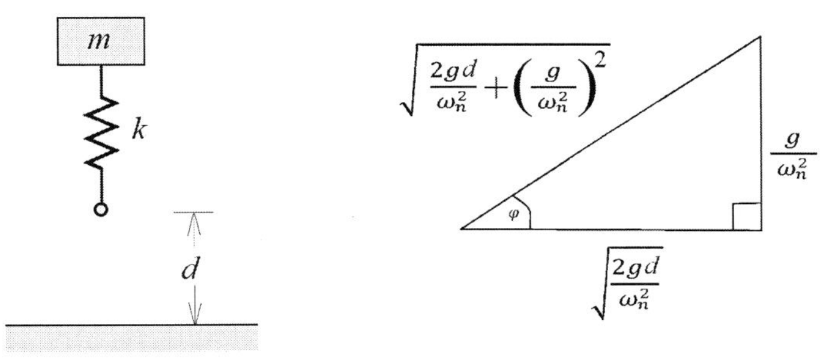

Figure 3.13: Spring-mass dropping through a vertical distance.

Assume $t = 0$ at the instant when the spring just touches the ground. With reference to Figure 3.13, one has the equation of motion as

$$m\ddot{x} + kx = mg\,, \qquad (3.26)$$

where g is the acceleration due to gravity.

In the above system the initial conditions are: $x(0) = 0$ and $\dot{x}(0) = \sqrt{2gd}$. One may use the Laplace transform method for the solution of the response since the rhs of Equation (3.26) is a constant. Thus, taking the Laplace transform of Equation (3.26), one has

$$m[s^2X(s) - sx(0) - \dot{x}(0)] + kX(s) = \frac{mg}{s}.$$

Substituting for the initial conditions, one obtain

$$s^2X(s) - \sqrt{2gd} + \frac{k}{m}X(s) = \frac{g}{s}.$$

Rearranging terms to give

$$X(s)(s^2 + \omega_n^2) = \sqrt{2gd} + \frac{g}{s}.$$

Therefore, the subsidiary equation becomes

$$X(s) = \frac{\sqrt{2gd}}{s^2 + \omega_n^2} + \frac{g}{s(s^2 + \omega_n^2)}. \tag{3.27}$$

Taking the inverse Laplace transform of $X(s)$, one has

$$x(t) = \frac{\sqrt{2gd}}{\omega_n} sin\omega_n t + \frac{g}{\omega_n^2}(1 - cos\omega_n t).$$

By making use of the right-angle triangle relationship shown in Figure 3.13 and the trigonometric identity

$$sin(\omega_n t - \varphi) = sin\,\omega_n t\ cos\varphi - cos\,\omega_n t\ sin\,\varphi,$$

one can show that the displacement becomes

$$x(t) = \frac{g}{\omega_n^2} + \sqrt{\frac{2gd}{\omega_n^2} + \left(\frac{g}{\omega_n^2}\right)^2}\ sin(\omega_n t - \varphi), \qquad x(t) > 0. \tag{3.28}$$

This equation agrees with that given by Thomson (Thomson, 1993, 99)

3.5 Questions and Solutions

In this section representative questions and their solutions are presented. The single dof systems considered are: (a) system involves rotational and translational motions, (b) system with rigid bar and lumped masses in translational motion, (c) massless rigid rod with spring and mass, (d) cylinder motion, (e) simplified model of a spring-supported vehicle, and (f) motion of water level meter.

Question 1
Show that the system shown in Figure 1Q1 is a single-degree-of-freedom (sdof) system. Determine the natural frequency of this system.

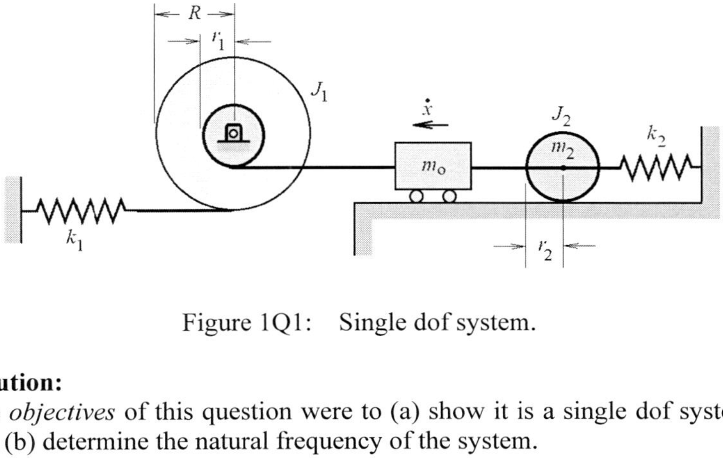

Figure 1Q1: Single dof system.

Solution:
The *objectives* of this question were to (a) show it is a single dof system, and (b) determine the natural frequency of the system.

It is assumed that oscillation is small.
In this problem there is no damping and therefore the energy method can be employed. It is concerned with general motion and therefore it involves with rotation and translation.

Let θ_1 and θ_2 be the small angular displacements of J_1 and J_2, respectively. Since J_1 has only rotational motion, m_o has only translation, while J_2 has translational as well as rotational motions such that the kinetic energy T for the system becomes

$$T = \frac{1}{2} J_1(\dot{\theta}_1)^2 + \frac{1}{2} m_o(\dot{x})^2 + \left[\frac{1}{2} m_2(\dot{x})^2 + \frac{1}{2} J_2(\dot{\theta}_2)^2\right].$$

for a small oscillation, $x = \theta_1 r_1 = \theta_2 r_2$ so that $\dot{x} = \dot{\theta}_1 r_1 = \dot{\theta}_2 r_2$ and

$$T = \frac{1}{2}\left(\frac{J_1}{r_1^2} + m_o + m_2 + \frac{J_2}{r_2^2}\right)(\dot{x})^2 = \frac{1}{2}\,M(\dot{x})^2$$

where $M = \frac{J_1}{r_1^2} + m_o + m_2 + \frac{J_2}{r_2^2}$ is the effective mass of the system.

Now, consider the potential energy U of the system. With reference to the figure, the potential energy is

$$U = \frac{1}{2}\,k_1(R\theta_1)^2 + \frac{1}{2}\,k_2(r_2\theta_2)^2$$

$$= \frac{1}{2}\,k_1\left(R\frac{x}{r_1}\right)^2 + \frac{1}{2}\,k_2\left(r_2\,\frac{x}{r_2}\right)^2 = U = Kx^2$$

in which the effective or equivalent stiffness of the system is

$$K = k_1\left(\frac{R}{r_1}\right)^2 + k_2.$$

Applying the principle of conservation of energy, that is, $\frac{d(T+U)}{dt} = 0$, one obtains

$$M\ddot{x} + Kx = 0$$

and therefore, the natural frequency of the system becomes

$$\omega_n = \sqrt{\frac{K}{M}} = \sqrt{\frac{k_1\left(\frac{R}{r_1}\right)^2 + k_2}{\frac{J_1}{r_1^2} + m_o + m_2 + \frac{J_2}{r_2^2}}}.$$

Question 2

Find the equivalent mass of the system shown in Figure 1Q2 in which the rigid bar has a mass moment of inertia J_o about the pin-joint (Hint: this is a sdof system).

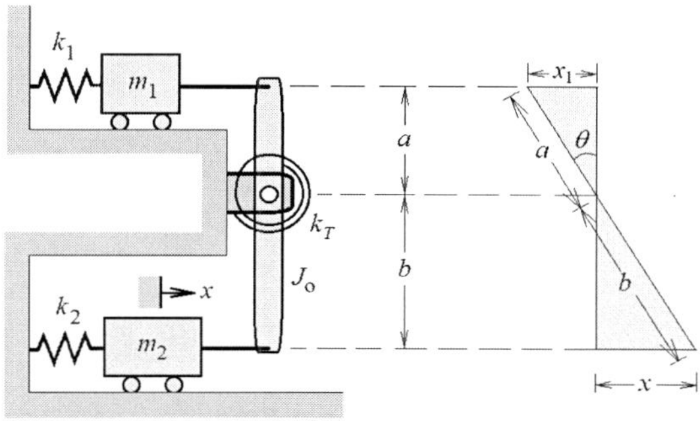

Figure 1Q2: Single dof system.

Solution:
It is assumed that oscillation is small. In this problem there is no damping and therefore the energy method can be employed.

It is concerned with general motion and therefore it involves with rotation and translation. Let θ and x_1 be the small angular displacement of the bar and displacement of mass m_1, respectively. Then $x = b\theta$ and $x_1 = a\theta$ or $x_1 = \left(\frac{a}{b}\right)x$.

The kinetic energy T of the system is

$$T = \frac{1}{2}m_1(\dot{x}_1)^2 + \frac{1}{2}m_2(\dot{x})^2 + \frac{1}{2}J_o\left(\dot{\theta}\right)^2$$

$$= \frac{1}{2}m_1\left(\frac{a}{b}\right)^2(\dot{x})^2 + \frac{1}{2}m_2(\dot{x})^2 + \frac{1}{2}J_o\left(\frac{\dot{x}}{b}\right)^2 = \frac{1}{2}M(\dot{x})^2$$

where the effective mass of the system is

$$M = m_1\left(\frac{a}{b}\right)^2 + m_2 + \frac{J_o}{b^2}.$$

Question 3
The horizontal massless rigid rod of length L is hinged at the left-hand side (lhs) and a concentrated mass m is attached to the right-hand side (rhs), as shown in Figure 1Q3. A spring of constant coefficient k and a damper of coefficient of c are attached at a distance a from the lhs. Derive the equation of motion for the system shown in the figure. With the derived equation of motion determine (a) the critical damping coefficient, and (b) the natural

frequency of damped oscillation.

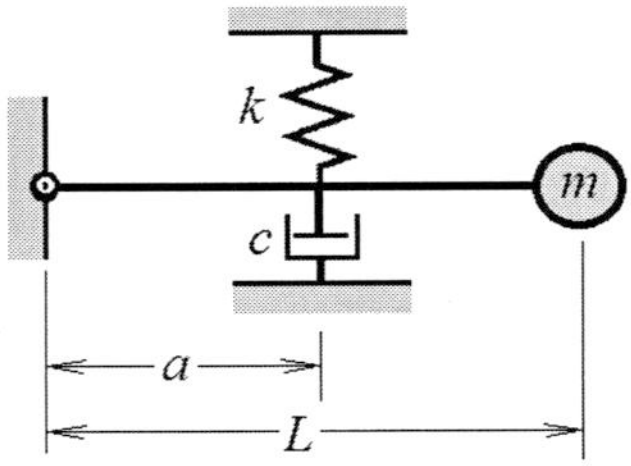

Figure 1Q3: Massless rigid rod with a damper and spring.

Solution:
The FBD of the system in Figure 1Q3 is included in Figure 1Q3a.

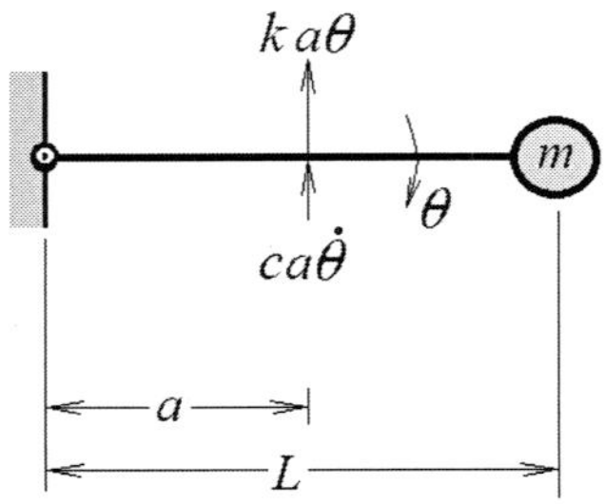

Figure 1Q3a: FBD of massless rigid rod with a damper and spring.

Applying the FBD in the above figure and taking moments about the hinge,

$$\sum M = m\big(L\ddot{\theta}\big)L = \; -a[c\big(a\dot{\theta}\big) - \; a[k(a\theta)]$$

which upon moving all terms to the lhs of the equality sign,

$$m\big(L\ddot{\theta}\big)L + a[c\big(a\dot{\theta}\big)] + \; a[k(a\theta)] = 0.$$

This equation of motion can be written as

$$\ddot{\theta}+\left(\frac{c}{m}\right)\left(\frac{a}{L}\right)^2\dot{\theta}+\left(\frac{k}{m}\right)\left(\frac{a}{L}\right)^2\theta=0. \tag{3.i}$$

Let $\theta = e^{st}$ such that $\dot{\theta} = se^{st}$ and $\ddot{\theta} = s^2e^{st}$.

Substituting these results into the last equation, one obtains the characteristic equation

$$s^2+\left(\frac{c}{m}\right)\left(\frac{a}{L}\right)^2 s+\left(\frac{k}{m}\right)\left(\frac{a}{L}\right)^2=0$$

whose roots are given by

$$s_{1,2}=-\frac{1}{2}\left(\frac{c}{m}\right)\left(\frac{a}{L}\right)^2\pm\frac{1}{2}\sqrt{\left[\left(\frac{c}{m}\right)\left(\frac{a}{L}\right)^2\right]^2-4\left(\frac{k}{m}\right)\left(\frac{a}{L}\right)^2}.$$

Note that in general $4\left(\frac{k}{m}\right)\left(\frac{a}{L}\right)^2>\left[\left(\frac{c}{m}\right)\left(\frac{a}{L}\right)^2\right]^2$ which means that system is lightly damped and the terms inside the square root sign is negative. Therefore, the above roots can be written as

$$s_{1,2}=-\frac{1}{2}\left(\frac{c}{m}\right)\left(\frac{a}{L}\right)^2\pm i\left(\frac{1}{2}\right)\sqrt{4\left(\frac{k}{m}\right)\left(\frac{a}{L}\right)^2-\left[\left(\frac{c}{m}\right)\left(\frac{a}{L}\right)^2\right]^2}. \tag{3.ii}$$

(a) For the critically damped case, the motion has no oscillatory feature.

That is, the imaginary part becomes zero or $4\left(\frac{k}{m}\right)\left(\frac{a}{L}\right)^2=\left[\left(\frac{c}{m}\right)\left(\frac{a}{L}\right)^2\right]^2$. Note that under this critical condition the damping coefficient c is written as c_c.Thus, $\left[\left(\frac{c_c}{m}\right)\left(\frac{a}{L}\right)^2\right]^2=4\left(\frac{k}{m}\right)\left(\frac{a}{L}\right)^2$ which leads to the critical damping coefficient, $c_c=(2L/a)\sqrt{km}$.

(b) The damped natural frequency or natural frequency of damped oscillation is

$$\omega_d=\omega_n\sqrt{1-\zeta^2}$$

in which the damping coefficient ζ is given by the equation of motion

$$\ddot{\theta} + 2\zeta\omega_n \dot{\theta} + \omega_n^2\theta = 0$$

where $2\zeta\omega_n = \left(\frac{c}{m}\right)\left(\frac{a}{L}\right)^2$ and $\omega_n^2 = \left(\frac{k}{m}\right)\left(\frac{a}{L}\right)^2$ so that

$$\omega_n = \left(\frac{a}{L}\right)\sqrt{\frac{k}{m}} \text{ and } \zeta = \frac{1}{2\omega_n}\left(\frac{c}{m}\right)\left(\frac{a}{L}\right)^2 \text{ or } \zeta^2 = \frac{1}{4\omega_n^2}\left[\left(\frac{c}{m}\right)\left(\frac{a}{L}\right)^2\right]^2.$$

The last term can be simplified as

$$\zeta^2 = \frac{1}{4\left(\frac{k}{m}\right)\left(\frac{a}{L}\right)^2}\left[\left(\frac{c}{m}\right)\left(\frac{a}{L}\right)^2\right]^2 = \frac{1}{4}\left(\frac{c}{m}\right)^2\left(\frac{a}{L}\right)^2\left(\frac{m}{k}\right)$$
$$= \frac{1}{4}\left(\frac{c^2}{km}\right)\left(\frac{a}{L}\right)^2.$$

Substituting into the damped natural frequency, one obtains

$$\omega_d = \left(\frac{a}{L}\right)\sqrt{\frac{k}{m}}\sqrt{1 - \frac{1}{4}\left(\frac{c^2}{km}\right)\left(\frac{a}{L}\right)^2}.$$

Question 4

A cylinder is shown in Figure 1Q4 in which the mass of the cylinder is *m* and is connected to a spring of stiffness *k* excited through viscous friction *c* to a piston displacement $y = Y\sin\omega t$. Determine the amplitude *X* of the cylinder motion *x* and its phase with respect to the piston.

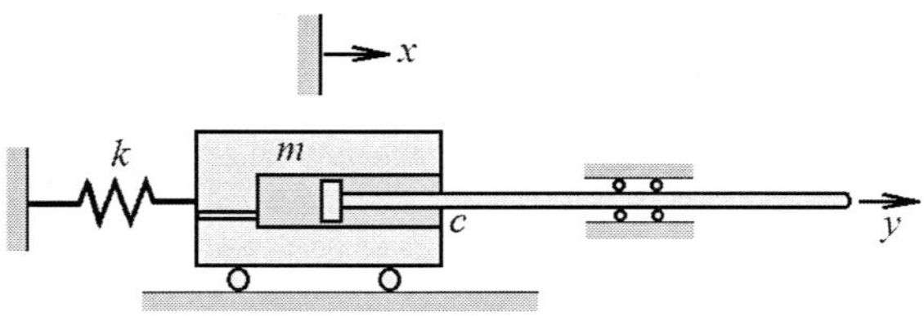

Figure 1Q4: Cylinder in motion.

Solution:
The FBD of the system in Figure 1Q4 is included in Figure 1Q4a. Applying the FBD in the above figure, the equation of motion can be obtained as

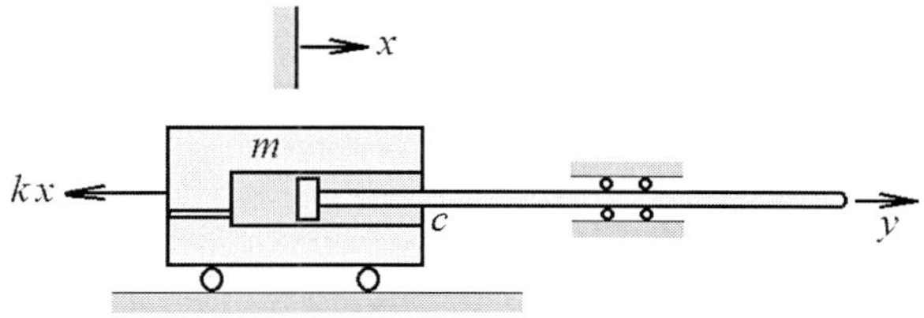

Figure 1Q4a: FBD of Cylinder in motion.

$$m\ddot{x} = -\,c(\dot{x} - \dot{y}) - kx.$$

Rearranging terms, one has

$$m\ddot{x} + c\dot{x} + kx = c\dot{y}\,.$$

Since $y = Ysin\,\omega$ one can write $y = Ye^{i\omega t}$ Thus, $\dot{y} = i\omega\, Ye^{i\omega t}$. Therefore, one can assume the response to be

$$x = Xe^{i(\omega t-\varphi)} = Xe^{-i\varphi}\,e^{i\omega t} = \bar{X}e^{i\omega t}, \qquad \bar{X} = Xe^{-i\varphi},$$

where φ is the phase angle due to the positive damping.

Substituting the above results into the equation of motion and after rearranging terms, one obtains

$$(k - m\omega^2 + ic\omega)\bar{X} = ic\omega Y.$$

Therefore,

$$\bar{X} = \frac{ic\omega Y}{k - m\omega^2 + ic\omega} \qquad \text{or} \qquad Xe^{-i\varphi} = \frac{ic\omega Y}{k - m\omega^2 + ic\omega}.$$

The rhs can be written as (recall $z = x + iy$ can be written as $z = \sqrt{x^2 + y^2}\,e^{i\mu}$),

$$\frac{ic\omega Y}{k - m\omega^2 + ic\omega} = \frac{ic\omega Ye^{-i\mu}}{\sqrt{(k - m\omega^2)^2 + (c\omega)^2}}$$

in which

$$\mu = tan^{-1}\left(\frac{c\omega}{k - m\omega^2}\right). \qquad (4.i)$$

Therefore,

$$Xe^{-i\varphi} = \frac{ic\omega Y}{k - m\omega^2 + ic\omega} = \frac{ic\omega Y e^{-i\mu}}{\sqrt{(k - m\omega^2)^2 + (c\omega)^2}} \qquad (4.ii)$$

That is, the amplitude X is given by

$$X = \frac{c\omega Y}{\sqrt{(k - m\omega^2)^2 + (c\omega)^2}}$$

and phase angle is determined the following

$$ie^{-i\mu} = e^{-i\mu + \frac{\pi}{2}}$$

which makes use of the fact that the product of an imaginary number and a vector is equal to the vector with it being rotated 90° counter-clockwise. One can also prove this as the following

$$ie^{-i\mu} = i\,(cos\mu - isin\mu) = sin\mu + icos\mu = e^{-i\left(\mu - \frac{\pi}{2}\right)}$$

because $e^{-i\left(\mu - \frac{\pi}{2}\right)} = cos\left(\mu - \frac{\pi}{2}\right) - isin\left(\mu - \frac{\pi}{2}\right)$

$$= cos\mu\, cos\frac{\pi}{2} + sin\mu\, sin\frac{\pi}{2} - \, i\left(\sin\mu\; cos\frac{\pi}{2} - cos\mu\, sin\frac{\pi}{2}\right)$$

$$= sin\mu + \, i\, cos\mu.$$

That is, $\varphi = \mu - \frac{\pi}{2}$. Collecting results, the amplitude

$$X = \frac{c\omega Y}{\sqrt{(k - m\omega^2)^2 + (c\omega)^2}}$$

and the phase angle w. r. t. the piston is $\mu - \frac{\pi}{2}$ where μ is given by Equation (4.*i*).

Question 5

A simplified model of a spring-supported vehicle traveling over a rough road (which is considered as a sinusoidal wave of amplitude *Y*) is shown in Figure 1Q5. (a) Derive the equation for the amplitude of the vehicle of weight $W = mg$ as a function of the speed *v* of the traveling vehicle, and (b) determine the most unfavorable speed (*Hint*: the most unfavorable speed corresponds to the resonance situation in which $\omega = \omega_n$).

Solution:

(a) Applying the FBD in Figure 1Q5, the equation of motion becomes

$$m\ddot{x} = -\,k(x-y) \quad \text{or} \quad m\ddot{x} + kx = ky.$$

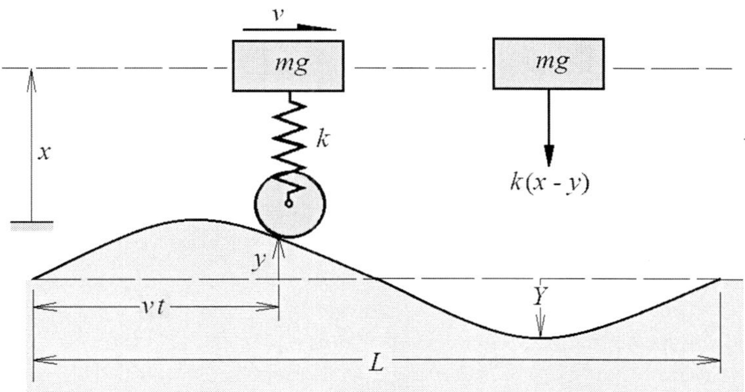

Figure 1Q5: Simplified model of a spring-supported vehicle.

(b) Since the rough road shown in the figure is represented as a sinusoidal form such that one can write

$$y = Y sin\,\omega t$$

where $\omega = \frac{2\pi}{\tau}$ with the period τ being related to the velocity of the vehicle v and wave length *L* by $v = \frac{L}{\tau}$. Thus,

$$\omega = \frac{2\pi}{L/v} = \frac{2\pi v}{L}.$$

Therefore,

$$y = Y sin\left(\frac{2\pi v}{L}t\right).$$

Since there is no damping and therefore the response has no delay such that one can assume $x = Xsin\ \omega t$.

Upon application of Equation (3.10), one obtains

$$X = \frac{Yk}{k - m\omega^2} = \frac{Y}{1 - \left(\frac{\omega}{\omega_n}\right)^2}$$

since $\omega_n^2 = \frac{k}{m}$.

The most unfavorable speed corresponds to the resonance situation. That is, when $\omega = \omega_n$ which leads to $X = \infty$. Thus, at resonance

$$\omega = 2\pi v/L = \sqrt{k/m} \quad \rightarrow \quad v = [L/(2\pi)]\sqrt{k/m} = [L/(2\pi)]\sqrt{kg/W}.$$

Therefore, the most unfavorable speed is $v = \frac{L}{2\pi}\sqrt{\frac{k}{m}}$.

Question 6

A water level meter, as shown in Figure 1Q6, composes of a massless rigid rod *OA* where a floating cylinder of diameter $2r$ and mass m is attached via a massless taut rod *AB*. Given that the density of water is ρ and acceleration due to gravity g, derive the equation of motion for the system shown.

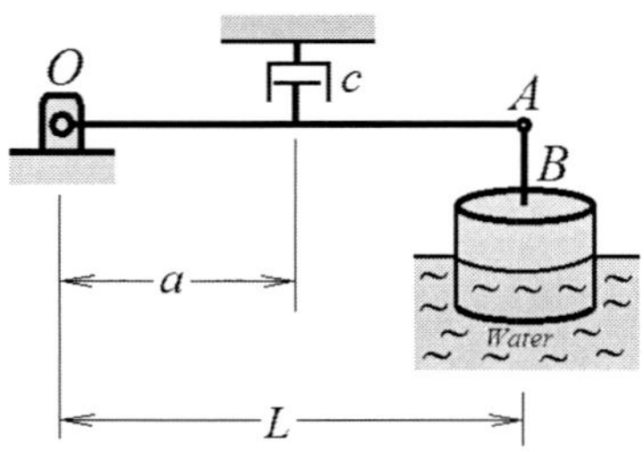

Figure 1Q6: A water level meter.

Solution:
Applying the FBD in Figure 1Q6a and taking moments about the hinge on the lhs,

$$\sum M = m(L\ddot{\theta})L = -a[c(a\dot{\theta})] - L[k(L\theta)]$$

where $k(L\theta)$ is the restoring force related to the weight of water displaced (recall, Archimedes' principle) such that

$$k(L\theta) = \rho(\pi r^2)(L\theta)g \qquad \text{or} \qquad k = \rho(\pi r^2)g.$$

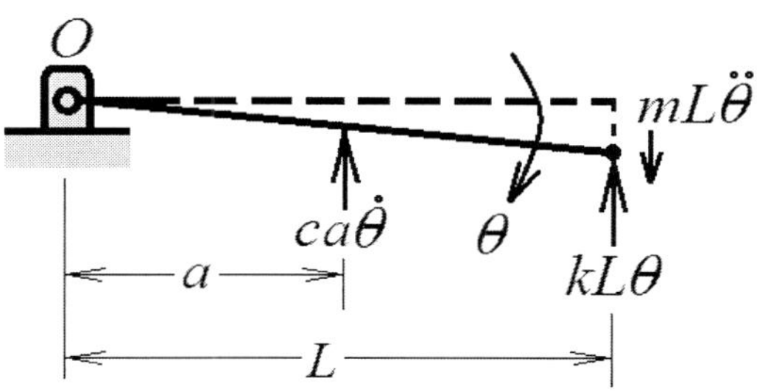

Figure 1Q6a: FBD of water level meter.

Therefore,

$$m(L\ddot{\theta})L + a[c(a\dot{\theta})] + L[\rho(\pi r^2)g(L\theta)] = 0.$$

This equation of motion can be written as

$$\ddot{\theta} + \left(\frac{c}{m}\right)\left(\frac{a}{L}\right)^2\dot{\theta} + \left(\frac{\rho g \pi r^2}{m}\right)\theta = 0.$$

Question 7
A step force is applied to a damped single dof system and its equation of motion is given by

$$\ddot{x} + 2\zeta\omega_n\dot{x} + \omega_n^2 x = g$$

where g is a constant. If the system starts from rest determine the displacement solution to this system.

Solution:
The general solution is

$$x = e^{-\zeta\omega_n t}(A_1 e^{i\omega_d t} + A_2 e^{-i\omega_d t}) + \frac{g}{\omega_d}\int_0^t e^{-\zeta\omega_n(t-\tau)}\sin[\omega_d(t-\tau)]d\tau$$

where the arbitrary constants A_1 and A_2 are obtained by applying the initial conditions. As the system starts from rest, these two constants are found to be zero. Therefore, the displacement solution is

$$x = \frac{g}{\omega_d}\int_0^t e^{-\zeta\omega_n(t-\tau)}\sin[\omega_d(t-\tau)]d\tau.$$

The Duhamel integral can be obtained by applying the formula in item 14.518 of (Spiegel, 1968, 85) which is

$$\int e^{ax}\sin bx\,dx = \frac{e^{ax}(a\sin bx - b\cos bx)}{a^2+b^2}$$

in which $a = -\zeta\omega_n$, $b = \omega_d$, $x = t-\tau$ such that $dx = -d\tau$.

Upon substituting, operating on the integral, and simplifying, one finds

$$\int_0^t e^{-\zeta\omega_n(t-\tau)}\sin\omega_d(t-\tau)\,d\tau = \left|\frac{e^{-\zeta\omega_n(t-\tau)}}{\omega_n^2}[\omega_d\cos\omega_d(t-\tau) + \zeta\omega_n\sin\omega_d(t-\tau)]\right|_0^t = \frac{\omega_d}{\omega_n^2} - \frac{e^{-\zeta\omega_n t}}{\omega_n^2}[\omega_d\cos\omega_d t + \zeta\omega_n\sin\omega_d t].$$

Therefore, the response becomes

$$x = \frac{g}{\omega_n^2}\left[1 - e^{-\zeta\omega_n t}\left(\cos\omega_d t + \frac{\zeta\omega_n}{\omega_d}\sin\omega_d t\right)\right].$$

References

Seto, W. W. 1964. Theory and Problems of Mechanical Vibrations. New York: McGraw-Hill.
Spiegel, M. R. 1968. Mathematical Handbook of Formulas and Tables. New York: McGraw-Hill.
Thomson, W. T. 1993. Theory of Vibration with Applications, Fourth Edition. Englewood Cliffs, New Jersey: Prentice-Hall.

Exercise Questions

Q5. A cylinder of radius r and mass m rolls without slipping and damping on a cylindrical surface of radius R as shown in Figure 3Q5. The motion is restricted to a plane vertical to the axis of the cylinder. If the cylinder is displaced from its equilibrium position, set up the differential equations for its motion and determine the natural frequency of small oscillations.

If now the cylinder is replaced by a sphere of same radius and same mass as those of the cylinder and it rolls without slipping and damping on a spherical surface of radius R similar to that shown in Figure 3Q5, set up the differential equations for its motion and determine the natural frequency of small oscillations. Moreover, in the above two cases which has a higher natural frequency?

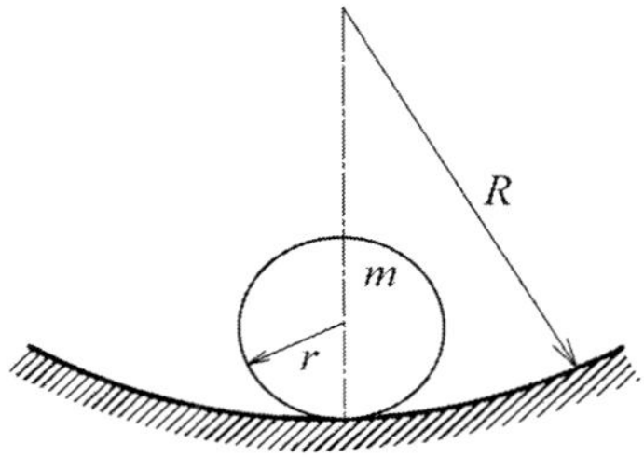

Figure 3Q5: A cylinder rolls without slipping and damping on a cylindrical surface.

Q6. A uniform rigid slender rod of mass m and length L is hinged at the middle. Both ends of the rod are attached by springs every one of which has a spring coefficient k as shown in Figure 3Q6. Assuming the rod is free to rotate about an axis perpendicular to the plane of the figure and through the hinge, find the natural frequency of small oscillation of the system.

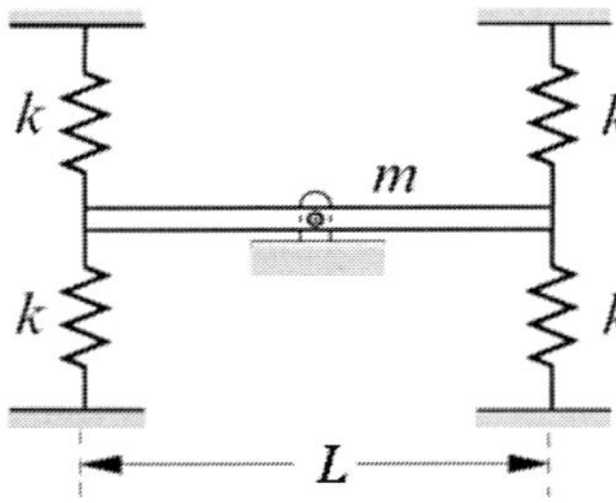

Figure 3Q6: Slender rod and springs arranged as a ‘H’ plane frame.

Q7. The single degree-of-freedom system shown in Figure 3.4 is excited by a force of periodic square wave as shown in Figure 3Q7. Determine the steady-state response x of the system.

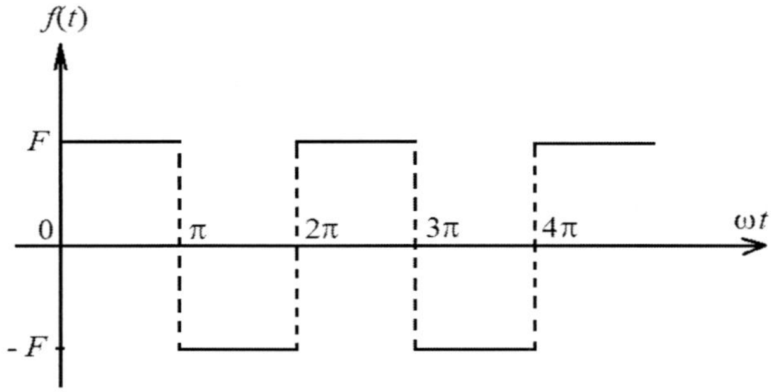

Figure 3Q7: Periodic square wave force.

Q8. A helicopter of mass m and its landing gear system represented as a spring constant of k tries to land on a rigid ground. However, owing to the downdraft it falls from a distance h before it touches the ground. The single degree-of-freedom model is shown in Figure 3Q8. Determine the time elapsed from the first contact of the landing gear system until it breaks contact again. Find the time when $h = \frac{mg}{4k}$ and its implication.

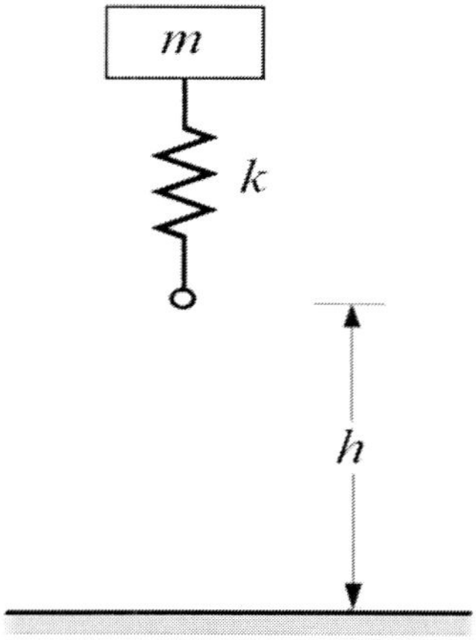

Figure 3Q8: A simplified landing helicopter model.

CHAPTER 4

FREE AND FORCED VIBRATION OF TWO DEGREES OF FREEDOM SYSTEMS

For many situations in which a first insight is provided by a simple model such as a single degree of freedom (sdof) system, however, this treatment may have to be refined in order to provide a more accurate solution. Moreover, there are various physical systems which are so complicated that a sdof model is not adequate. Therefore, a two degree of freedom (tdof) system may be considered. In this chapter the focus is on the analysis of linear tdof systems. In Section 4.1 the formulation and derivation of equations of motion for free vibration and *modal analysis* of a tdof system are presented. Section 4.2 is concerned with the forced harmonic vibration of a tdof system. The application of this tdof system to passive vibration control in which the so-called *dynamic absorber* is employed, is given in Section 4.3. Coordinate coupling is introduced in Section 4.4. In Section 4.5 normal mode vibration and responses of tdof system are included. Normal mode vibration of a tdof system with proportional damping is presented in Section 4.6. The final section in this chapter, Section 4.7 deals with the forced vibration of a tdof system with non-proportional damping.

4.1 Free Vibration and Modal Analysis of Two Degrees of Freedom Systems

Free vibration of tdof systems belongs to the special case of the class of free vibration of many degrees (or multi-degree) of freedom (mdof) systems. In order to provide approaches in dealing with tdof systems while preserving concise treatment for the topics included in this chapter, in this section only two tdof systems are included for illustration. These two physical systems are the coal sizing machine and expensive electronic equipment installed inside a rocket. They are sufficiently representative to provide an appreciation of how to translate them from the physical world into the models of the conceptual world. The three stages are outlined below.

Stage 1: ***Systems in Physical World***

The coal sizing machine essentially consists of a jig and an absorber mass connected by springs to the frame, as shown in Figure 4.1 in which the screen reciprocates with a frequency. This reciprocating action is the source of harmonic force applied to the jig. The second physical world example is the expensive electronic equipment installed inside the rocket. The electronic equipment is supported by a frame structure that can be modelled as massless beams, as sketched in Figure 4.1.

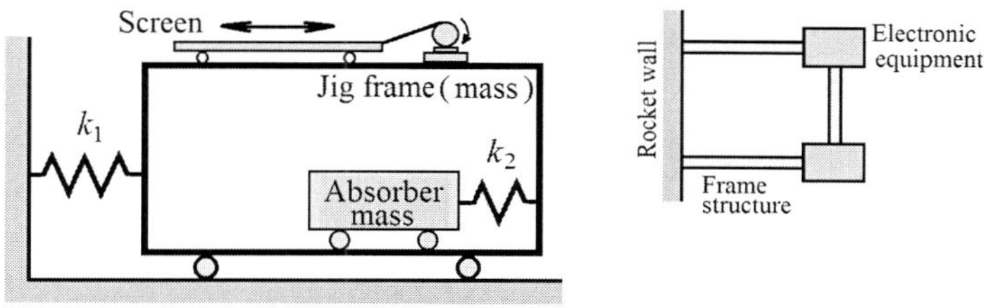

Figure 4.1: Two tdof systems in physical world.

Stage 2: ***Systems in Conceptual World***

The two tdof systems in the physical world may be analysed as the lumped-parameter models in the conceptual or mathematical world. They are presented in Figure 4.2.

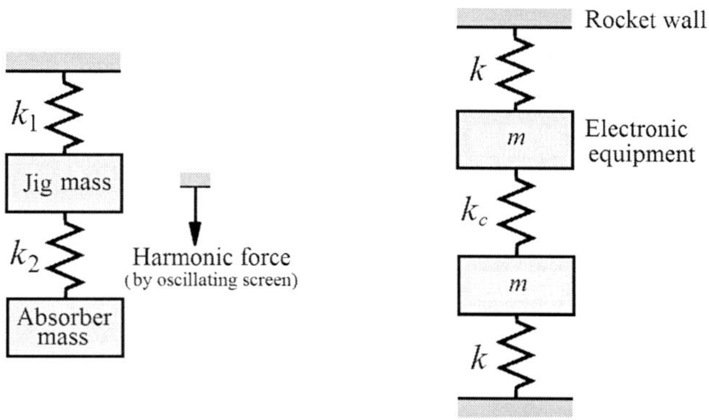

Figure 4.2: Lumped-parameter models.

Stage 3: ***Formulation and Solution***

Once the conceptual model is established one may proceed with the derivation of the governing equation of motion for the individual system. For brevity, only the coal sizing machine is analysed in details in the following. Consider the lumped-parameter model of this system and the free-body diagram (FBD) of the tdof system, as shown in Figure 4.3.

For simplicity damping in the system is disregarded so that by applying Newton's law of motion to the jig mass, one has

$$m_1\ddot{x}_1 = \sum F = -k_1x_1 - k_2(x_1 - x_2) + F_1 sin\omega t.$$

Rearranging terms, it becomes

$$m_1\ddot{x}_1 + (k_1 + k_2)x_1 - k_2x_2 = F_1 sin\omega t.$$

Similarly, applying Newton's second law of motion for the absorber or second mass and the FBD, the equation of motion is given by

$$m_2\ddot{x}_2 = -k_2(x_2 - x_1)$$

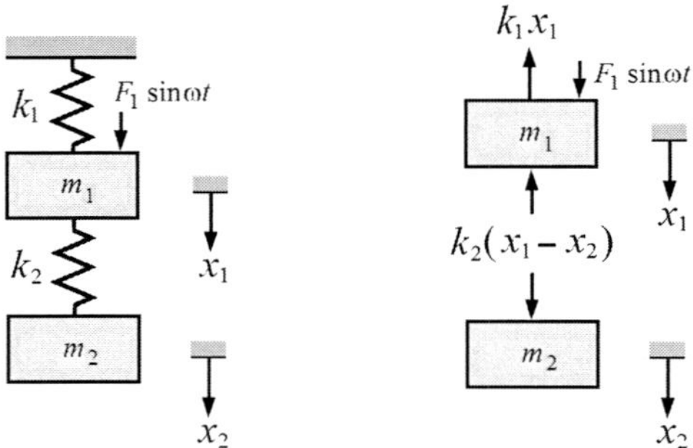

Figure 4.3: Coal sizing machine model (left) and its FBD (right).

or after rearranging terms

$$m_2\ddot{x}_2 + k_2(x_2 - x_1) = 0\,.$$

For subsequent analysis, one can express the above two equations of motion in a more compact form. To this end, matrix is applied. Thus, the above two equations of motion can be written in matrix form as

$$\begin{bmatrix} m_1 & 0 \\ 0 & m_2 \end{bmatrix} \begin{pmatrix} \ddot{x}_1 \\ \ddot{x}_2 \end{pmatrix} + \begin{bmatrix} (k_1 + k_2) & -k_2 \\ -k_2 & k_2 \end{bmatrix} \begin{pmatrix} x_1 \\ x_2 \end{pmatrix} = \begin{pmatrix} F_1 \\ 0 \end{pmatrix} \sin \omega t.$$

This equation can be written mathematically as

$$\begin{bmatrix} m_{11} & 0 \\ 0 & m_{22} \end{bmatrix} \begin{pmatrix} \ddot{x}_1 \\ \ddot{x}_2 \end{pmatrix} + \begin{bmatrix} k_{11} & k_{12} \\ k_{21} & k_{22} \end{bmatrix} \begin{pmatrix} x_1 \\ x_2 \end{pmatrix} = \begin{pmatrix} F_1 \\ 0 \end{pmatrix} sin\omega t, \tag{4.1}$$

where $m_{11} = m_1$, $m_{22} = m_2$, $k_{11} = k_1 + k_2$, $k_{12} = k_{21} = -k_2$, $k_{22} = k_2$. When the rhs is set to zero this resulting equation is that for *free vibration*. The latter will be considered later in the present section.

Meanwhile, note that the diagonal elements or entries of the mass and stiffness matrices are all positive. Since the system has no damping and therefore the responses have no delay or phase difference with respect to the applied force, one can write

$$\begin{pmatrix} x_1 \\ x_2 \end{pmatrix} = \begin{pmatrix} X_1 \\ X_2 \end{pmatrix} \sin \omega t \,. \tag{4.2}$$

In this equation, X_i (i = 1, 2) are the amplitudes of instantaneous displacements x_i and symbolically not to be confused with the Laplace transform of x_i in Chapter 3.

Substituting Eq. (4.2) and it derivatives into Eq. (4.1) and upon simplification, one has

$$\begin{bmatrix} (k_{11} - m_1\omega^2) & k_{12} \\ k_{21} & (k_{22} - m_2\omega^2) \end{bmatrix} \begin{pmatrix} X_1 \\ X_2 \end{pmatrix} = \begin{pmatrix} F_1 \\ 0 \end{pmatrix} \tag{4.3}$$

or writing it in a more concise form

$$[\,Z(\omega)] \begin{pmatrix} X_1 \\ X_2 \end{pmatrix} = \begin{pmatrix} F_1 \\ 0 \end{pmatrix}$$

where the so-called *impedance matrix* is

$$[\,Z(\omega)] = \begin{bmatrix} (k_{11} - m_1\omega^2) & k_{12} \\ k_{21} & (k_{22} - m_2\omega^2) \end{bmatrix}.$$

Pre-multiplying on both sides of Eq. (4.3) by the inverse of $[Z(\omega)]$, it becomes

$$\binom{X_1}{X_2} = [\,Z(\omega)]^{-1}\binom{F_1}{0} = \frac{adj[\,Z(\omega)]\binom{F_1}{0}}{|Z(\omega)|} \qquad \text{or}$$

$$\binom{X_1}{X_2} = \frac{\begin{bmatrix}(k_{22} - m_2\omega^2) & -k_{12} \\ -k_{21} & (k_{11} - m_1\omega^2)\end{bmatrix}\binom{F_1}{0}}{|Z(\omega)|}\,. \tag{4.4a}$$

The determinant in the denominator on the rhs of Eq. (4.4*a*) may be expressed

$$|Z(\omega)| = m_1 m_2\,(\omega_1^2 - \omega^2)(\omega_2^2 - \omega^2) \tag{4.4b}$$

where ω_1 and ω_2 are the natural or normal mode frequencies of the tdof system.

- The proof of Eq. (4. 4*b*) and the expressions for the two natural frequencies are provided in Appendix 4A.

Applying Eq. (4.4*b*), one can write the elements of the amplitude displacement vector of Eq. (4.4*a*), respectively as

$$X_1 = \frac{(k_{22} - m_2\omega^2)F_1}{m_1 m_2\,(\omega_1^2 - \omega^2)(\omega_2^2 - \omega^2)}\,, \tag{4.5a}$$

$$X_2 = \frac{-\,k_{21}F_1}{m_1 m_2\,(\omega_1^2 - \omega^2)(\omega_2^2 - \omega^2)}\,. \tag{4.5b}$$

The above equations are very useful, in practice, for the elimination of vibration in machines and structural systems. To illustrate their use, the so-called *dynamic absorber* system is presented in the next section.

Now, attention is directed at the so-called *modal analysis* for tdof linear systems. One of the main reasons for performing such a modal analysis is that the system of coupled second order differential equations of motion can be uncoupled such that every uncoupled second order differential equation of motion may be simply analysed similar to that for the sdof system. After all the uncoupled equations of motion are solved they can be transformed back to the original coordinate system.

When there is no applied force Eq. (4.1) reduces to two homogeneous second order differential equations. After application of Eq. (4.2), one obtains Eq. (4.3) with the rhs reduces to a zero column vector. This is the case of free vibration. Thus, Eq. (4.3) now becomes

$$\begin{bmatrix} (k_{11} - m_1\omega^2) & k_{12} \\ k_{21} & (k_{22} - m_2\omega^2) \end{bmatrix} \begin{pmatrix} X_1 \\ X_2 \end{pmatrix} = \begin{pmatrix} 0 \\ 0 \end{pmatrix}. \tag{4.6}$$

The problem finding the values of the parameter $\lambda = \omega^2$ for which Eq. (4.6) admits nontrivial solutions is called the *characteristic-value problem*, or the *eigenvalue problem*. The *characteristic* or *frequency equation* of the system is obtained by setting the determinant of the impedance matrix to zero. That is,

$$\begin{vmatrix} (k_{11} - m_1\omega^2) & k_{12} \\ k_{21} & (k_{22} - m_2\omega^2) \end{vmatrix} = 0.$$

The physical basis for the fact that the two natural frequencies are obtained from the frequency equation when the determinant of the impedance matrix becomes zero is because at either one of these frequencies the amplitudes of displacements of the tdof system approach infinity, meaning the system is in resonance. The above frequency equation can be written as

$$a\lambda^2 + b\lambda + c = 0$$

where $a = m_1 m_2$, $b = -(k_{11}m_2 + k_{22}m_1)$, and $c = k_{11}k_{22} - k_{12}k_{21}$. Solving the quadratic equation one finds

$$\lambda_1 = \frac{-b + \sqrt{b^2 - 4ac}}{2a}, \quad \lambda_2 = \frac{-b - \sqrt{b^2 - 4ac}}{2a}. \tag{4.7}$$

These two roots of the quadratic equation are called the *eigenvalues* of the system. Note that $\lambda_1 = \omega_1^2$ and $\lambda_2 = \omega_2^2$, where ω_1^2 and ω_2^2 are the squares of the two natural frequencies defined by Eq. (*A.2a*) and (*A.2b*) in Appendix 4A, respectively.

After obtaining the eigenvalues, one substitutes λ_1 into Eq. (4.6) to find the amplitude ratio as

$$\left(\frac{X_1}{X_2}\right)_{(1)} = \frac{k_{12}}{m_1\omega_1^2 - k_{11}} = \frac{m_2\omega_1^2 - k_{22}}{k_{21}}, \tag{4.8a}$$

where the subscript (1) on the lhs of Eq. (4.8*a*) denotes the amplitude ratio associated with the first natural frequency.

Similarly, substituting λ_2 into Eq. (4.6), one has

$$\left(\frac{X_1}{X_2}\right)_{(2)} = \frac{k_{12}}{m_1\omega_2^2 - k_{11}} = \frac{m_2\omega_2^2 - k_{22}}{k_{21}}. \tag{4.8b}$$

Note that Eq. (4.8) only enables one to obtain the amplitude ratios and not their absolute values. In practice or numerical computation, one frequently selects unity or any other numerical value as one of the amplitudes so that it is referred to as the amplitude ratio being *normalized* to that value. This normalized amplitude ratio is called the *normal mode* or *natural mode*. Mathematically, it is known as the *eigenvector*. Thus, the normal modes are regarded as the steady state vibration patterns or *configurations* of the system at their corresponding natural frequencies.

Before leaving this section it should be noted that different natural modes are *orthogonal with respect to the mass and stiffness matrices* of the system. This is known as *orthogonality* or *orthogonality condition* which, for example, means

$$\left(\begin{pmatrix} \tilde{X}_1 \\ \tilde{X}_2 \end{pmatrix}_{(1)}\right)^T \begin{bmatrix} m_{11} & 0 \\ 0 & m_{22} \end{bmatrix} \begin{pmatrix} \tilde{X}_1 \\ \tilde{X}_2 \end{pmatrix}_{(2)} = 0, \tag{4.9}$$

in which $\begin{pmatrix} \tilde{X}_1 \\ \tilde{X}_2 \end{pmatrix}_{(2)}$ is the normalized natural mode of $\begin{pmatrix} X_1 \\ X_2 \end{pmatrix}_{(2)}$ associated with the second natural frequency of the system. In Section 4.5 the steps in obtaining the normal modes for a specific tdof system and the orthogonality condition will be presented in more detail.

4.2 Forced Harmonic Vibration and Dynamic Absorber

By making use of Eq. (4.5*a*) and recall that the denominator on the rhs is the determinant of impedance matrix $[Z(\omega)]$, one has

$$X_1 = \frac{(k_{22} - m_2\omega^2)F_1}{\begin{vmatrix} (k_{11} - m_1\omega^2) & k_{12} \\ k_{21} & (k_{22} - m_2\omega^2) \end{vmatrix}} \quad \text{or}$$

$$X_1 = \frac{(k_{22} - m_2\omega^2)F_1}{(k_{11} - m_1\omega^2)(k_{22} - m_2\omega^2) - k_{12}k_{21}}. \tag{4.10a}$$

The numerator becomes zero when the following relation is satisfied

$$\omega^2 = \frac{k_{22}}{m_2}.$$

In other words, if m_1 is the vibrating machine and m_2 is the dynamic absorber, when the vibrating frequency of the machine is equal to the natural frequency of the dynamic absorber the amplitude of machine vibration is zero.

In general, it is more convenient to express the amplitudes of the displacements in dimensionless forms. To this end, one can write Eq. (4.10*a*) as

$$X_1 = \frac{k_{22}\left(1 - \frac{m_2}{k_{22}}\omega^2\right)F_1}{k_{11}k_{22}\left(1 - \frac{m_1}{k_{11}}\omega^2\right)\left(1 - \frac{m_2}{k_{22}}\omega^2\right) - k_{12}k_{21}}.$$

This equation can be non-dimensionalized by moving F_1 and k_{11} from the rhs to the lhs. Thus,

$$\frac{X_1 k_{11}}{F_1} = \frac{\left(1 - \frac{m_2}{k_{22}}\omega^2\right)}{\left(1 - \frac{m_1}{k_{11}}\omega^2\right)\left(1 - \frac{m_2}{k_{22}}\omega^2\right) - \frac{k_{12}k_{21}}{k_{11}k_{22}}}$$

This can be simplified by writing $\Omega_1^2 = \frac{k_{11}}{m_1}$, $\Omega_2^2 = \frac{k_{22}}{m_2}$ so that the dimensionless amplitude of the displacement of the vibrating machine becomes

$$\frac{X_1 k_{11}}{F_1} = \frac{1 - \left(\frac{\omega}{\Omega_2}\right)^2}{\left[1 - \left(\frac{\omega}{\Omega_1}\right)^2\right]\left[1 - \left(\frac{\omega}{\Omega_2}\right)^2\right] - \frac{k_{12}k_{21}}{k_{11}k_{22}}}. \qquad (4.10b)$$

Similarly, the dimensionless amplitude of displacement of the dynamic absorber is

$$\frac{X_2 k_{11}}{F_1} = \frac{-1}{\frac{k_{22}}{k_{12}}\left\{\left[1 - \left(\frac{\omega}{\Omega_1}\right)^2\right]\left[1 - \left(\frac{\omega}{\Omega_2}\right)^2\right] - \frac{k_{12}k_{21}}{k_{11}k_{22}}\right\}}.$$

Since in this tdof system $k_{22} = k_2 = -k_{12}$, therefore it can be further simplified to give

$$\frac{X_2 k_{11}}{F_1} = \frac{1}{\left[1-\left(\frac{\omega}{\Omega_1}\right)^2\right]\left[1-\left(\frac{\omega}{\Omega_2}\right)^2\right]-\frac{k_2}{k_{11}}}. \qquad (4.10c)$$

- Note that $\Omega_1 \neq \omega_1$ and $\Omega_2 \neq \omega_2$. The natural frequencies ω_1 and ω_2 of the tdof system are derived in Appendix 4A.

4.3 Forced Harmonic Vibration of Tdof System with Viscous Damping

When the tdof system with viscous damping is harmonically forced it is relatively simple to obtain its steady-state displacements by employing the *mechanical impedance method* in which vectors are represented by complex numbers. To illustrate the mechanical impedance method the following tdof system is considered.

Example 4.1

A centrifugal pump is supported on a rigid foundation of mass m_2 through isolator springs of stiffness k_1. This centrifugal pump has an unbalance of *me.* The tdof system is shown in Figure 4.4. If the soil stiffness and damping are respectively, k_2 and c_2, find the displacements of the pump and foundation for the system in which

m = 0.25 kg, e = 0.1525 m (or 15.25 cm), k_1 = 230 MN/m, k_2 = 115 MN/m,

m_1 = 364 kg, m_2 = 909 kg, c_2 = 0.25 MNs/m, and

the speed of the pump is 1200 rpm.

Solution:

(a) ***Given parameters***

Given that the speed of the pump is 1200 rpm, therefore, the applied frequency is $\omega = 2\pi(1200)/60$ rad/s = 125.66 rad/s.

The applied force due to unbalance,

$$f_1 = me\omega^2 \sin \omega t$$
$$= 0.25\,(0.1525)(125.66)^2 \sin 125.66t \quad \text{N.}$$
$$= 602 \sin 125.66t \quad \text{N.}$$

The remaining given parameters are:

$$k_1 = 230\,\frac{\mathrm{MN}}{\mathrm{m}} = 2.3\times 10^8\,\frac{\mathrm{N}}{\mathrm{m}}, \qquad m_1 = 364\,\mathrm{kg},$$
$$k_2 = 115\,\mathrm{MN/m} = 1.15\times 10^8\,\mathrm{N/m}, \quad m_2 = 909\,\mathrm{kg},$$
$$c_2 = 0.25\ \mathrm{MNs/m} = 0.25\times 10^6\ \mathrm{MNs/m}.$$

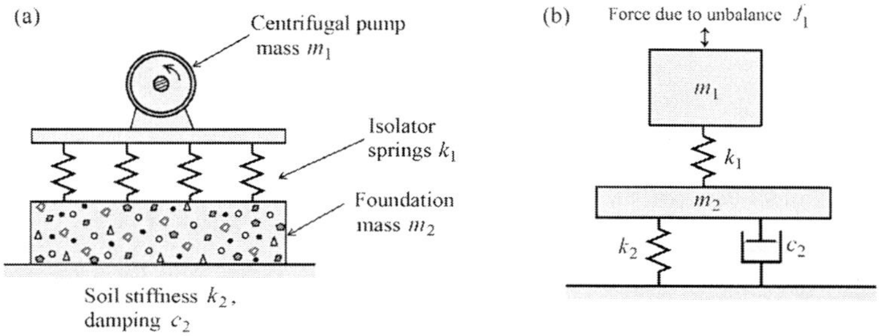

Figure 4.4: (a) Unbalanced centrifugal pump on foundation, and (b) conceptual model.

(b) ***Equations of motion***

Let x_1 and x_2 be the absolute equilibrium vertical displacements of the centrifugal pump (not including the unbalance mass *m*) and foundation, respectively.

With reference to Figure 4.5(a), the vertical displacement of *m* is

$$x_1 + e\ \sin\omega t\,.$$

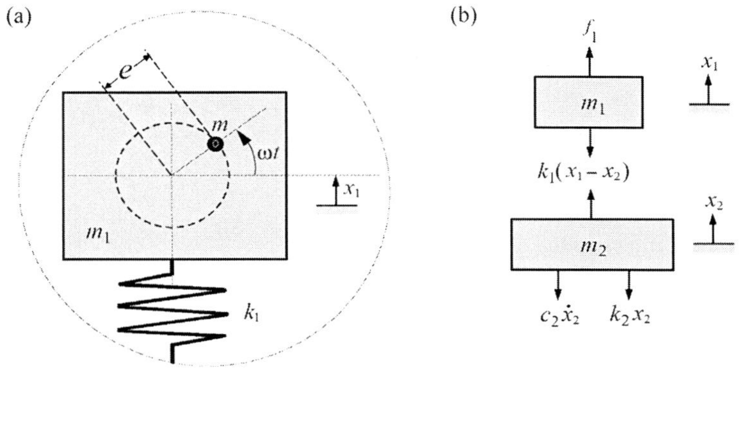

Figure 4.5: (a) Kinematics of unbalance mass, and (b) FBD of system.

With reference to this equation and the FBD in Figure 4.5(b), by applying Newton's second law of motion for the centrifugal pump of mass m_1 and the unbalance mass m, one has

$$m_1\ddot{x}_1 + m\,\frac{d^2(x_1 + e\,\sin\omega t\,)}{dt^2} = \sum F = -\,k_1(x_1 - x_2).$$

Rearranging terms, it becomes

$$(m_1 + m)\ddot{x}_1 + k_1x_1 - k_1x_2 = f_1 = F_1\sin\omega t\,, \qquad F_1 = me\omega^2.$$

As the unbalance mass m is much smaller than m_1 one can simply writes

$$m_1\ddot{x}_1 + k_1x_1 - k_1x_2 = f_1 = F_1\sin\omega t\,.$$

Similarly, applying Newton's second law of motion for the foundation m_2, the equation of motion is given by

$$m_2\ddot{x}_2 = -\,c_2\dot{x}_2 - k_2x_2 + k_1(x_1 - x_2).$$

After rearranging terms, it leads to

$$m_2\ddot{x}_2 + c_2\dot{x}_2 - k_1x_1 + x_2(k_1 + k_2) = 0\,.$$

For subsequent analysis, one can express the two equations of motion in matrix form. Thus, the above two equations of motion can be written as

$$\begin{bmatrix} m_1 & 0 \\ 0 & m_2 \end{bmatrix}\begin{pmatrix} \ddot{x}_1 \\ \ddot{x}_2 \end{pmatrix} + \begin{pmatrix} 0 & 0 \\ 0 & c_2 \end{pmatrix}\begin{pmatrix} \dot{x}_1 \\ \dot{x}_2 \end{pmatrix} + \begin{bmatrix} k_1 & -\,k_1 \\ -\,k_1 & k_{22} \end{bmatrix}\begin{pmatrix} x_1 \\ x_2 \end{pmatrix} = \begin{pmatrix} F_1 \\ 0 \end{pmatrix}\sin\omega t \quad (i)$$

in which $k_{22} = k_1 + k_2$.

(c) ***Solution for responses***

For convenience, one can write $\sin\omega t$ as $Im\{e^{i\omega t}\}$ such that $x_i = X_i\sin\omega t\,, i = 1, 2,$ can be written as $x_i = Im\{X_ie^{i\omega t}\}$, where $Im\{.\}$ denotes the imaginary part of the enclosing complex variable.

Without loss of generality, one can disregard the symbol $Im\{.\}$ for the time being so that the displacements can be written as $x_i = X_ie^{i\omega t}$. Then, the velocity and acceleration are given by $\dot{x}_i = i\omega\,X_ie^{i\omega t}$, $\ddot{x}_i = -\,\omega^2X_ie^{i\omega t}$. Substituting the above expressions into Eq. (i) and upon cancelling the Euler expression $e^{i\omega t}$ on both sides of the resulting matrix equation of motion, one has

$$\begin{bmatrix} -\omega^2 m_1 & 0 \\ 0 & -\omega^2 m_2 \end{bmatrix}\begin{pmatrix} X_1 \\ X_2 \end{pmatrix} + \begin{pmatrix} 0 & 0 \\ 0 & i\omega c_2 \end{pmatrix}\begin{pmatrix} X_1 \\ X_2 \end{pmatrix} + \begin{bmatrix} k_1 & -k_1 \\ -k_1 & k_{22} \end{bmatrix}\begin{pmatrix} X_1 \\ X_2 \end{pmatrix} = \begin{pmatrix} F_1 \\ 0 \end{pmatrix}.$$

Upon simplification, one arrives at

$$\begin{bmatrix} k_1 - \omega^2 m_1 & -k_1 \\ -k_1 & k_1 + k_2 - \omega^2 m_2 + i\omega c_2 \end{bmatrix}\begin{pmatrix} X_1 \\ X_2 \end{pmatrix} = \begin{pmatrix} F_1 \\ 0 \end{pmatrix}. \qquad (ii)$$

This is identical in form to Eq. (4.3). Thus, one can write the coefficient matrix as

$$[Z(\omega)] = \begin{bmatrix} k_1 - \omega^2 m_1 & -k_1 \\ -k_1 & k_1 + k_2 - \omega^2 m_2 + i\omega c_2 \end{bmatrix}. \qquad (iii)$$

The form of the impedance or coefficient matrix $[Z(\omega)]$ is identical to that in Eq. (4.3) for a tdof system. That is, the elements of the coefficient matrix are $z_{11} = k_1 - \omega^2 m_1$, $z_{12} = z_{21} = -k_1$, $z_{22} = k_1 + k_2 - \omega^2 m_2 + i\omega c_2$.

From Eq. (4.7), the amplitudes of the displacements are

$$X_1 = \frac{z_{22} F_1}{|Z(\omega)|} = \frac{z_{22} F_1}{z_{11}\, z_{22} - z_{12} z_{21}}.$$

Therefore,

$$X_1 = \frac{k_1 F_1}{(k_1 - \omega^2 m_1)(k_1 + k_2 - \omega^2 m_2 + i\omega c_2) - k_1^2}, \qquad (iv)$$

and

$$X_2 = \frac{-z_{12} F_1}{|Z(\omega)|} = \frac{(k_1 + k_2 - \omega^2 m_2 + i\omega c_2) F_1}{(k_1 - \omega^2 m_1)(k_1 + k_2 - \omega^2 m_2 + i\omega c_2) - k_1^2}. \qquad (v)$$

Now, substituting the given data into the numerator and denominator terms in Eq. (iv),

$$k_1 + k_2 - \omega^2 m_2 + i\omega c_2$$

$$= [(2.3 + 1.15) \times 10^8 - 125.66^2(909) + i(125.66 \times 0.25) \times 10^6]602 \quad \text{N/m}$$

$$= 19.904919 \times 10^{10} + i(1.891183) \times 10^{10} \quad \text{N/m}$$

$$(k_1 - \omega^2 m_1)(k_1 + k_2 - \omega^2 m_2 + i\omega c_2) - k_1^2$$

$$= 7.4148238 \times 10^{16} - 5.29 \times 10^{16} + i(0.7044886) \times 10^{16} \quad (\text{N/m})^2$$

$$= (2.1248238 + i0.7044886) \times 10^{16} \quad (\text{N/m})^2$$

Therefore,

$$X_1 = \frac{19.904919 \times 10^{10} + i(1.891183) \times 10^{10}}{(2.1248238 + i0.7044886) \times 10^{16}} \text{ m.}$$

Rationalization, one has

$$X_1 = \frac{(19.904919 + i1.891183)(2.1248238 - i0.7044886) \times 10^{-6}}{(2.1248238 + i0.7044886)(2.1248238 - i0.7044886)} \text{ m.}$$

$$= \frac{(43.5801 - i\,10.0044) \times 10^{-6}}{2.1248238^2 + 0.7044886^2}\, m$$

$$= (8.6870 - i\,1.9942) \times 10^{-6} \quad \text{m}.$$

From the theory of complex variables,

$$X_1 = (8.6870 - i\,1.9942) \times 10^{-6} \quad \text{m} = X_{10} e^{i\varphi_1} \quad \text{m},$$

where

$$X_{10} = \sqrt{8.6870^2 + 1.9942^2} \times 10^{-6} \quad \text{m}, \qquad \varphi_1 = tan^{-1}\left(\frac{-\,1.9942}{8.6870}\right).$$

Therefore,

$$X_{10} = 8.9129 \times 10^{-6} \quad \text{m}, \qquad \varphi_1 = -12.9^o.$$

Thus,

$$x_1 = Im\{X_1 e^{i\omega t}\} = Im\{X_{10} e^{i(\omega t + \varphi_1)}\} = X_{10}\sin(\omega t + \varphi_1).$$

$$= 8.9129 \times 10^{-6}\sin(\omega t - 12.9^o) \quad \text{m.} \qquad (vi)$$

Similarly, applying Eq.(v), one obtains

$$X_2 = \frac{13.8460 \times 10^{-6}}{(2.1248238 + i0.7044886)} \text{ m.}$$

Rationalization, one has

$$X_2 = \frac{13.8460\,(2.1248238 - i0.7044886) \times 10^{-6}}{(2.1248238 + i0.7044886)(2.1248238 - i0.7044886)} \text{ m.}$$

$$= (5.8660 - i\,1.9449) \times 10^{-6} \text{ m.}$$

$$= (5.8660 - i\,1.9449) \times 10^{-6} \quad \text{m} = X_{20} e^{i\varphi_2} \quad \text{m},$$

where

$$X_{20} = \sqrt{5.8660^2 + 1.9942^2} \times 10^{-6} \quad \text{m}, \qquad \varphi_2 = tan^{-1}\left(\frac{-1.9942}{5.8660}\right).$$

Therefore,

$$X_{20} = 6.18 \times 10^{-6} \quad \text{m}, \qquad \varphi_2 = -18.3^o .$$

Thus,

$$x_2 = Im\{X_2 e^{i\omega t}\} = Im\{X_{20} e^{i(\omega t + \varphi_2)}\} = X_{20}\sin(\omega t + \varphi_2).$$

$$= 6.1800 \times 10^{-6}\sin(\omega t - 18.3^o) \quad \text{m.} \qquad (vii)$$

4.4 Co-ordinate Coupling

As shown in the foregoing every tdof system has two coupled second order differential equations. In general, there are three types of coupling in tdof and mdof systems. When the generalized coordinate x (vertical translation and transverse to the length of the rigid bar) is taken at the center of gravity G of the rigid uniform bar supported at both ends by springs, as shown in Figure 4.6 in which the tdof system experiences *elastic* or *static coupling.*

Mathematically, this is the case in which the off-diagonal elements of the stiffness are not zero. The second type of couplings is the so-called *inertia* or *dynamic coupling* in which the off-diagonal elements of the mass matrix are not zero. This occurs if the generalized coordinate x is taken at some point away from the center of gravity G but not at either end of the bar. The third type is the *static and dynamic coupling* which arises when the generalized coordinate is taken at one end of the bar. In this type of coupling the off-diagonal elements of the mass and stiffness matrices are not zero.

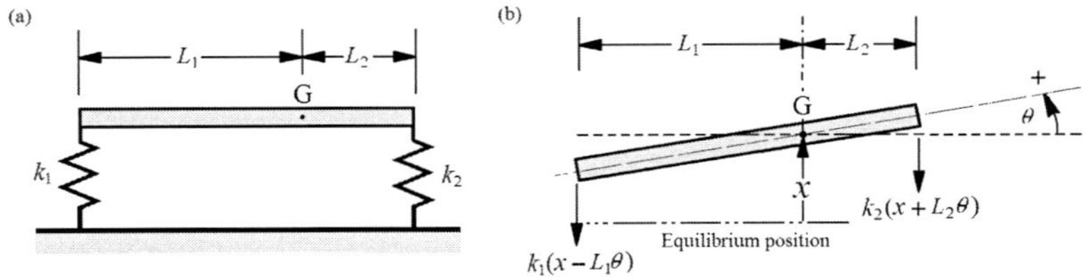

Figure 4.6: (a) Rigid uniform bar, and (b) FBD.

In this section, only the static coupling is considered since the other two types can easily and similarly be dealt with.

The rigid uniform bar supported at both end by two springs, as shown in Figure 4.6, is displaced vertically upward and the angular rotation θ is assumed to be small. By Newton's second law of motion, summing forces vertically, one obtains

$$m\ddot{x} = \sum F = -k_1(x - L_1\theta) - k_2(x + L_2\theta).$$

Upon rearranging and moving all terms to the lhs of the equality sign, one has

$$m\ddot{x} + (k_1 + k_2)x - (k_1L_1 - k_2L_2)\theta = 0. \qquad (4.11a)$$

It may be appropriate to note that the weight mg of the bar does not appear in the equation of motion because the static deflections in the springs are supposed to be balanced by the weight. In other words, the translational motion x is dynamic while the static deflection due to the weight of the bar should not be included in the above equation of motion.

Now, by summing the moments about an axis through G, one obtains

$$J\ddot{\theta} = k_1(x - L_1\theta)L_1 - k_2(x + L_2\theta)L_2.$$

where J is the mass moment of inertia about G. Upon rearranging terms, one has

$$J\ddot{\theta} - (k_1L_1 - k_2L_2)x + (k_1L_1^2 + k_2L_2^2)\theta = 0. \quad (4.11b)$$

Writing in matrix form, Eqs. (4.11a) and (4.11b) give

$$\begin{bmatrix} m & 0 \\ 0 & J \end{bmatrix}\begin{pmatrix} \ddot{x} \\ \ddot{\theta} \end{pmatrix} + \begin{bmatrix} k_1 + k_2 & -(k_1L_1 - k_2L_2) \\ -(k_1L_1 - k_2L_2) & k_1L_1^2 + k_2L_2^2 \end{bmatrix}\begin{pmatrix} x \\ \theta \end{pmatrix} = \begin{pmatrix} 0 \\ 0 \end{pmatrix}. \quad (4.12)$$

When $k_1L_1 = k_2L_2$ the off-diagonal elements of the stiffness matrix become zero. Thus, the two second order differential equations are uncoupled. This means that the tdof system becomes two sdof systems. That is, the generalized coordinates x and θ are independent from each other.

4.5 Normal Mode Vibration of Forced Tdof Systems

The normal vibration of forced tdof systems is presented in this section. Consider a mass-spring system of tdof as shown in Figure 4.7. The objectives are to (a) derive the equations of motion and find the natural frequencies of the system, given that $m_1 = m_2 = m$, $k_1 = k_3 = 4k$, $k_2 = k$; and (b) find the responses of the system by making use of the modal or transformation matrix, assuming the system starts from rest and the force $f_1 = f_1(t) = \tilde{u}(t)$, where $\tilde{u}(t)$ is the unit step function.

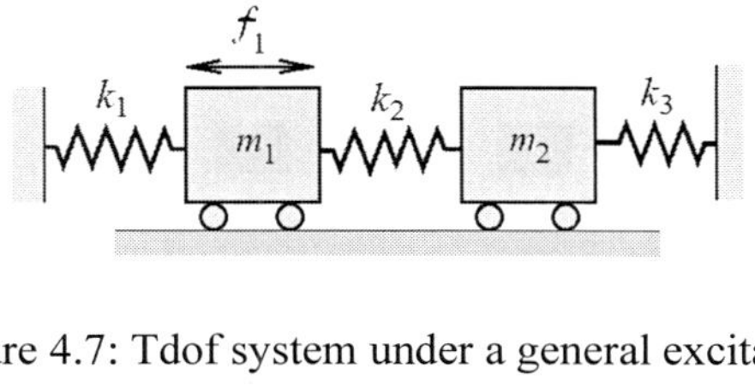

Figure 4.7: Tdof system under a general excitation.

The equations of motion can be obtained by applying Newton's second law of motion. With reference to the FBD in Figure 4.8, one obtains

$$m_1\ddot{x}_1 = -k_1x_1 - k_2(x_1 - x_2) + f_1,$$

$$m_2\ddot{x}_2 = -k_3x_2 + k_2(x_1 - x_2).$$

Rearranging terms,

$$m_1\ddot{x}_1 + (k_1 + k_2)x_1 - k_2x_2 = f_1,$$

$$m_2\ddot{x}_2 + (k_3 + k_2)x_2 - k_2x_1 = 0.$$

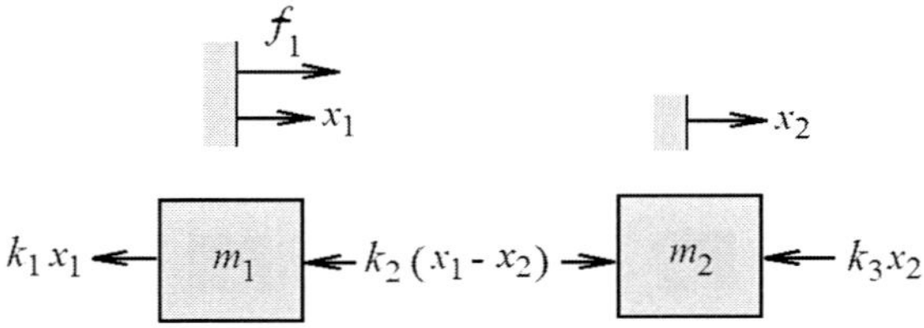

Figure 4.8: FBD of tdof system in Figure 4.6.

Solution:
(a) ***Equations of motion***
After substituting for the given data, it leads to

$$m\ddot{x}_1 + 5kx_1 - kx_2 = f_1(t), \tag{4.13a}$$

$$m\ddot{x}_2 + 5kx_2 - kx_1 = 0. \tag{4.13b}$$

Writing these equations in matrix form,

$$\begin{bmatrix} m & 0 \\ 0 & m \end{bmatrix}\begin{pmatrix} \ddot{x}_1 \\ \ddot{x}_2 \end{pmatrix} + \begin{bmatrix} 5k & -k \\ -k & 5k \end{bmatrix}\begin{pmatrix} x_1 \\ x_2 \end{pmatrix} = \begin{pmatrix} f_1(t) \\ 0 \end{pmatrix}. \tag{4.14}$$

This is the required equation of motion for the given system.

(b) ***Forced responses of system by modal analysis***
Since the system has no damping and for oscillatory motion one can assume

$$x_1 = X_1 \cos \omega t, \quad x_2 = X_2 \cos \omega t$$

so that

$$\ddot{x}_1 = -\omega^2 X_1 \cos \omega t, \quad \ddot{x}_2 = -\omega^2 X_2 \cos \omega t.$$

To find the modal matrix, one substitutes these values into Eq. (4.14) and disregarding the forcing vector, the resulting matrix equation becomes

$$\begin{bmatrix} (5k - m\omega^2) & -k \\ -k & (5k - m\omega^2) \end{bmatrix} \begin{pmatrix} X_1 \\ X_2 \end{pmatrix} = \begin{pmatrix} 0 \\ 0 \end{pmatrix}. \tag{4.15}$$

The frequency equation or characteristic equation is obtained by setting the determinant of the coefficient matrix to zero. That is,

$$\begin{vmatrix} (5k - m\omega^2) & -k \\ -k & (5k - m\omega^2) \end{vmatrix} = 0. \tag{4.16}$$

Operating the determinant, the frequency equation becomes

$$(5k - m\omega^2)^2 - k^2 = 0 \qquad \rightarrow \qquad m^2\omega^4 - (10km)\,\omega^2 + 24k^2 = 0$$

which gives the two natural frequencies for the system

$$\omega_1 = 2\sqrt{k/m}, \quad \omega_2 = \sqrt{6k/m}.$$

Natural modes or mode shapes or eigenvectors

To obtain the mode shapes of the system one considers Eq. (4.15),

$$\begin{bmatrix} (5k - m\omega^2) & -k \\ -k & (5k - m\omega^2) \end{bmatrix} \begin{pmatrix} X_1 \\ X_2 \end{pmatrix} = \begin{pmatrix} 0 \\ 0 \end{pmatrix}.$$

For a particular mode, one can write this equation as

$$\begin{bmatrix} (5k - m\omega_r^2) & -k \\ -k & (5k - m\omega_r^2) \end{bmatrix} \begin{pmatrix} X_{1r} \\ X_{2r} \end{pmatrix} = \begin{pmatrix} 0 \\ 0 \end{pmatrix}. \tag{4.17}$$

For $\omega_1^2 = 4\,\frac{k}{m}$, one can either apply the first or the second homogeneous equation and assume, for convenience, $X_{11} = 1$ to obtain

$$(5k - m\omega_1^2)X_{11} - kX_{21} = 0 \qquad \text{which gives} \quad X_{21} = 1.$$

Therefore, the first natural mode or mode shape can be written as

$$\begin{pmatrix} X_{11} \\ X_{21} \end{pmatrix} = \begin{pmatrix} X_1 \\ X_2 \end{pmatrix}_{(1)} = \begin{pmatrix} 1 \\ 1 \end{pmatrix}. \tag{4.18a}$$

Similarly, for $\omega_2^2 = 6\,\frac{k}{m}$ and assuming $X_{12} = 1$, the second natural mode becomes

$$(5k - m\omega_2^2)X_{12} - kX_{22} = 0 \qquad \text{which gives} \quad X_{22} = -1.$$

Therefore, the second natural mode or mode shape becomes

$$\begin{pmatrix} X_{12} \\ X_{22} \end{pmatrix} = \begin{pmatrix} X_1 \\ X_2 \end{pmatrix}_{(2)} = \begin{pmatrix} 1 \\ -1 \end{pmatrix}. \qquad (4.18b)$$

Note that in Eq. (4.18), the second subscript in the vector denotes the modal number.

Before proceeding further, one can normalize the natural modes. To this end, one let the normalized natural modes be

$$\begin{pmatrix} \tilde{X}_1 \\ \tilde{X}_2 \end{pmatrix}_{(1)} \text{ and } \begin{pmatrix} \tilde{X}_1 \\ \tilde{X}_2 \end{pmatrix}_{(2)} \text{ such that } \begin{pmatrix} \tilde{X}_1 \\ \tilde{X}_2 \end{pmatrix}_{(1)} = \mu_1 \begin{pmatrix} X_1 \\ X_2 \end{pmatrix}_{(1)} = \mu_1 \begin{pmatrix} 1 \\ 1 \end{pmatrix}_{(1)}$$

and

$$\begin{pmatrix} \tilde{X}_1 \\ \tilde{X}_2 \end{pmatrix}_{(2)} = \mu_2 \begin{pmatrix} X_1 \\ X_2 \end{pmatrix}_{(2)} = \mu_2 \begin{pmatrix} 1 \\ -1 \end{pmatrix}_{(2)}$$

where the constants μ_1 and μ_2 are to be found as in the following steps. For μ_1, one performs the matrix operation

$$\left(\begin{pmatrix} \tilde{X}_1 \\ \tilde{X}_2 \end{pmatrix}_{(1)}\right)^T \begin{bmatrix} m & 0 \\ 0 & m \end{bmatrix} \begin{pmatrix} \tilde{X}_1 \\ \tilde{X}_2 \end{pmatrix}_{(1)} = \mu_1 [X_1 \quad X_2]_{(1)} \begin{bmatrix} m & 0 \\ 0 & m \end{bmatrix} \mu_1 \begin{pmatrix} X_1 \\ X_2 \end{pmatrix}_{(1)} = 1$$

which gives

$$\mu_1^2 [1 \quad 1]_{(1)} \begin{bmatrix} m & 0 \\ 0 & m \end{bmatrix} \begin{pmatrix} 1 \\ 1 \end{pmatrix}_{(1)} = 1.$$

Operating on the matrix relation, one has $\mu_1^2(2m) = 1$ giving $\mu_1 = \frac{1}{\sqrt{2m}}$.

For μ_2, one performs the matrix operation

$$\left(\begin{pmatrix}\tilde{X}_1\\ \tilde{X}_2\end{pmatrix}_{(2)}\right)^T \begin{bmatrix} m & 0\\ 0 & m\end{bmatrix}\begin{pmatrix}\tilde{X}_1\\ \tilde{X}_2\end{pmatrix}_{(2)} = \mu_2[X_1 \quad X_2]_{(2)}\begin{bmatrix} m & 0\\ 0 & m\end{bmatrix}\mu_2\begin{pmatrix}X_1\\ X_2\end{pmatrix}_{(2)} = 1$$

which is

$$\mu_2^2[1 \quad -1]_{(2)}\begin{bmatrix} m & 0\\ 0 & m\end{bmatrix}\begin{pmatrix}1\\ -1\end{pmatrix}_{(2)} = 1.$$

Thus, $\mu_2^2(2m) = 1$ giving $\mu_2 = \frac{1}{\sqrt{2m}}$.

Therefore, the normalized natural modes are, respectively

$$\begin{pmatrix}\tilde{X}_1\\ \tilde{X}_2\end{pmatrix}_{(1)} = \mu_1\begin{pmatrix}X_1\\ X_2\end{pmatrix}_{(1)} = \frac{1}{\sqrt{2m}}\begin{pmatrix}1\\ 1\end{pmatrix}_{(1)} \quad \text{and}$$

$$\begin{pmatrix}\tilde{X}_1\\ \tilde{X}_2\end{pmatrix}_{(2)} = \mu_2\begin{pmatrix}X_1\\ X_2\end{pmatrix}_{(2)} = \frac{1}{\sqrt{2m}}\begin{pmatrix}1\\ -1\end{pmatrix}_{(2)}. \quad (4.19)$$

Test of orthogonality condition of natural modes

Consider the following,

$$\mu_1^2[1 \quad 1]_{(1)}\begin{bmatrix} m & 0\\ 0 & m\end{bmatrix}\begin{pmatrix}1\\ -1\end{pmatrix}_{(2)} = 0 \quad \text{and} \quad \mu_2^2[1 \quad -1]_{(2)}\begin{bmatrix} m & 0\\ 0 & m\end{bmatrix}\begin{pmatrix}1\\ 1\end{pmatrix}_{(1)} = 0.$$

Therefore, the orthogonality condition with respect to the mass matrix is satisfied. The proof of orthogonality condition with respect to the stiffness matrix is left to the reader.

Diagonalization of stiffness matrix

To obtain the diagonalized stiffness matrix, one is required first to form the modal or transformation matrix

$$[R] = \left[\begin{pmatrix}\tilde{X}_1\\ \tilde{X}_2\end{pmatrix}_{(1)} \quad \begin{pmatrix}\tilde{X}_1\\ \tilde{X}_2\end{pmatrix}_{(2)}\right] = \frac{1}{\sqrt{2m}}\begin{bmatrix}1 & 1\\ 1 & -1\end{bmatrix}. \quad (4.20)$$

Then, to obtain the diagonalized stiffness matrix, one performs

$$[R]^T\begin{bmatrix} 5k & -k \\ -k & 5k \end{bmatrix}[R]$$
$$=\frac{1}{\sqrt{2m}}\begin{bmatrix} 1 & 1 \\ 1 & -1 \end{bmatrix}\begin{bmatrix} 5k & -k \\ -k & 5k \end{bmatrix}\frac{1}{\sqrt{2m}}\begin{bmatrix} 1 & 1 \\ 1 & -1 \end{bmatrix}$$

$$=\frac{1}{2m}\begin{bmatrix} 1 & 1 \\ 1 & -1 \end{bmatrix}\begin{bmatrix} 4k & 6k \\ 4\,k & -6k \end{bmatrix}=\frac{k}{2m}\begin{bmatrix} 8 & 0 \\ 0 & 12 \end{bmatrix}=\frac{k}{m}\begin{bmatrix} 4 & 0 \\ 0 & 6 \end{bmatrix}.$$

This shows that

$$[R]^T\begin{bmatrix} 5k & -k \\ -k & 5k \end{bmatrix}[R]=\frac{k}{m}\begin{bmatrix} 4 & 0 \\ 0 & 6 \end{bmatrix}=\begin{bmatrix} \omega_1^2 & 0 \\ 0 & \omega_2^2 \end{bmatrix}.$$

Uncoupling the matrix equation of motion

Recall that the idea of solving the matrix equation of motion by the modal analysis is to make use of the solution of sdof system. This means that one has to uncouple the matrix equation of motion by first performing the coordinate transformation. To this end, one writes

$$\begin{pmatrix} x_1 \\ x_2 \end{pmatrix}=[R]\begin{pmatrix} q_1 \\ q_2 \end{pmatrix} \qquad \text{so that} \qquad \begin{pmatrix} \ddot{x}_1 \\ \ddot{x}_2 \end{pmatrix}=[R]\begin{pmatrix} \ddot{q}_1 \\ \ddot{q}_2 \end{pmatrix}. \quad (4.21a,b)$$

Substituting the above into Eq. (4.14), one obtains

$$\begin{bmatrix} m & 0 \\ 0 & m \end{bmatrix}[R]\begin{pmatrix} \ddot{q}_1 \\ \ddot{q}_2 \end{pmatrix}+\begin{bmatrix} 5k & -k \\ -k & 5k \end{bmatrix}[R]\begin{pmatrix} q_1 \\ q_2 \end{pmatrix}=\begin{pmatrix} f_1(t) \\ 0 \end{pmatrix}. \quad (4.22)$$

Pre-multiplying this equation by the transpose of the modal matrix,

$$[R]^T\begin{bmatrix} m & 0 \\ 0 & m \end{bmatrix}[R]\begin{pmatrix} \ddot{q}_1 \\ \ddot{q}_2 \end{pmatrix}+[R]^T\begin{bmatrix} 5k & -k \\ -k & 5k \end{bmatrix}[R]\begin{pmatrix} q_1 \\ q_2 \end{pmatrix}=[R]^T\begin{pmatrix} f_1(t) \\ 0 \end{pmatrix}. \quad (4.23)$$

But

$$[R]^T\begin{bmatrix} m & 0 \\ 0 & m \end{bmatrix}[R]=\begin{bmatrix} 1 & 0 \\ 0 & 1 \end{bmatrix}, \quad [R]^T\begin{bmatrix} 5k & -k \\ -k & 5k \end{bmatrix}[R]=\begin{bmatrix} \omega_1^2 & 0 \\ 0 & \omega_2^2 \end{bmatrix},$$

$$[R]^T=\frac{1}{\sqrt{2m}}\begin{bmatrix} 1 & 1 \\ 1 & -1 \end{bmatrix},$$

therefore, Eq. (4.23) reduces to

$$\begin{bmatrix} 1 & 0 \\ 0 & 1 \end{bmatrix} \begin{pmatrix} \ddot{q}_1 \\ \ddot{q}_2 \end{pmatrix} + \begin{bmatrix} \omega_1^2 & 0 \\ 0 & \omega_2^2 \end{bmatrix} \begin{pmatrix} q_1 \\ q_2 \end{pmatrix} = \frac{1}{\sqrt{2m}} \begin{pmatrix} f_1(t) \\ f_1(t) \end{pmatrix}. \quad (4.24)$$

That is, the uncoupled equations of motion are, respectively

$$\ddot{q}_1 + \omega_1^2 q_1 = \frac{1}{\sqrt{2m}} f_1(t) \quad \text{and} \quad \ddot{q}_2 + \omega_2^2 q_2 = \frac{1}{\sqrt{2m}} f_1(t).$$

Since the system starts from rest, the solutions to these two independent second order differential equations are, respectively

$$q_1 = \frac{1}{\sqrt{2m}} \left(\frac{1}{\omega_1}\right) \int_0^t f_1(\tau) \sin \omega_1 (t - \tau)\ d\tau \quad \text{and}$$

$$q_2 = \frac{1}{\sqrt{2m}} \left(\frac{1}{\omega_2}\right) \int_0^t f_1(\tau) \sin \omega_2 (t - \tau)\ d\tau.$$

Upon substituting for $f_1(t) = \tilde{u}(t)$, one obtains

$$\begin{aligned} q_1 &= \frac{1}{\sqrt{2m}} \left(\frac{1}{\omega_1}\right) \int_0^t \tilde{u}(\tau) \sin \omega_1 (t - \tau)\ d\tau \\ &= \frac{1}{\sqrt{2m}} \left(\frac{1}{\omega_1}\right) \int_0^t \sin \omega_1 (t - \tau)\ d\tau. \\ &= \frac{1}{\sqrt{2m}} \left(\frac{1}{\omega_1}\right) [-(-) \cos \omega_1 (t - \tau)]_0^t \\ &= \frac{1}{\sqrt{2m}} \left(\frac{1}{\omega_1^2}\right) (1 - \cos \omega_1 t), \end{aligned} \quad (4.25a)$$

$$\begin{aligned} q_2 &= \frac{1}{\sqrt{2m}} \left(\frac{1}{\omega_2}\right) \int_0^t \tilde{u}(\tau) \sin \omega_2 (t - \tau)\ d\tau \\ &= \frac{1}{\sqrt{2m}} \left(\frac{1}{\omega_2}\right) \int_0^t \sin \omega_2 (t - \tau)\ d\tau \\ &= \frac{1}{\sqrt{2m}} \left(\frac{1}{\omega_2^2}\right) (1 - \cos \omega_2 t). \end{aligned} \quad (4.25b)$$

Responses of system

The responses of the system, according to Eq. (4.21*a*) become

$$\begin{pmatrix} x_1 \\ x_2 \end{pmatrix} = [R]\begin{pmatrix} q_1 \\ q_2 \end{pmatrix} = \frac{1}{\sqrt{2m}}\begin{bmatrix} 1 & 1 \\ 1 & -1 \end{bmatrix}\begin{pmatrix} q_1 \\ q_2 \end{pmatrix}$$

which, upon expansion, gives

$$x_1 = \frac{q_1}{\sqrt{2m}} + \frac{q_2}{\sqrt{2m}} = \frac{1}{2m}\left(\frac{1}{\omega_1^2}\right)(1 - \cos\omega_1 t) + \frac{1}{2m}\left(\frac{1}{\omega_2^2}\right)(1 - \cos\omega_2 t)$$

$$x_1 = \frac{1}{8k}\left[1 - \cos\left(2\sqrt{k/m}\,t\right)\right] + \frac{1}{12k}\left[1 - \cos\left(\sqrt{6k/m}\,t\right)\right], \quad (4.26a)$$

$$x_2 = \frac{1}{8k}\left[1 - \cos\left(2\sqrt{k/m}\,t\right)\right] - \frac{1}{12k}\left[1 - \cos\left(\sqrt{6k/m}\,t\right)\right]. \quad (4.26b)$$

4.6 Normal Mode Vibration of Forced Tdof Systems with Proportional Damping

The system considered in the last section has no damping. Therefore, it is natural to include *proportional* or *Rayleigh damping* in the same system. By making use of Eq. (4.14) in Section 4.5, the matrix equation of motion now becomes

$$\begin{bmatrix} m & 0 \\ 0 & m \end{bmatrix}\begin{pmatrix} \ddot{x}_1 \\ \ddot{x}_2 \end{pmatrix} + [C]\begin{pmatrix} \dot{x}_1 \\ \dot{x}_2 \end{pmatrix} + \begin{bmatrix} 5k & -k \\ -k & 5k \end{bmatrix}\begin{pmatrix} x_1 \\ x_2 \end{pmatrix} = \begin{pmatrix} f_1(t) \\ 0 \end{pmatrix}, \quad (4.27)$$

where the proportional damping matrix is defined as

$$[C] = \lambda_m[M] + \lambda_k[K] = \lambda_m\begin{bmatrix} m & 0 \\ 0 & m \end{bmatrix} + \lambda_k\begin{bmatrix} 5k & -k \\ -k & 5k \end{bmatrix} \quad (4.28)$$

in which the constants λ_m and λ_k are determined by the two specified damping ratios and natural frequencies of the system. The steps in obtaining these constants will be included later. Meanwhile, going through the steps in obtaining the uncoupled equations for the undamped system as in Section 4.5, one obtains

$$\begin{bmatrix}1 & 0\\0 & 1\end{bmatrix}\begin{pmatrix}\ddot{q}_1\\\ddot{q}_2\end{pmatrix} + \left(\lambda_m\begin{bmatrix}1 & 0\\0 & 1\end{bmatrix} + \lambda_k\begin{bmatrix}\omega_1^2 & 0\\0 & \omega_2^2\end{bmatrix}\right)\begin{pmatrix}\dot{q}_1\\\dot{q}_2\end{pmatrix}$$

$$+\begin{bmatrix}\omega_1^2 & 0\\0 & \omega_2^2\end{bmatrix}\begin{pmatrix}q_1\\q_2\end{pmatrix} = \frac{1}{\sqrt{2m}}\begin{pmatrix}f_1(t)\\f_1(t)\end{pmatrix}. \qquad (4.29)$$

That is, the individual uncoupled equation of motion is

$$\ddot{q}_r + (\lambda_m + \lambda_k\omega_r^2)\dot{q}_r + \omega_r^2 q_r = \frac{f_r(t)}{\sqrt{2m}}, \qquad r = 1, 2. \quad (4.30)$$

Note that according to Eq. (4.14), in this particular system, $f_2(t) = f_1(t)$.

By applying the Duhamel integral and the fact that the initial conditions are zero, the solution to Eq. (4.30) is

$$q_r = \frac{1}{\sqrt{2m}}\left(\frac{1}{\omega_{dr}}\right)\int_0^t e^{-\zeta_r\omega_r(t-\tau)}\, f_r(\tau)\, \sin\omega_{dr}(t-\tau)\, d\tau \qquad (4.31)$$

where the damped natural frequency ω_{dr} and damping ratio or modal damping parameter ζ_r for mode r are defined as

$$\omega_{dr} = \omega_r\sqrt{1-\zeta_r^2}, \qquad 2\zeta_r\omega_r = \lambda_m + \lambda_k\omega_r^2. \qquad (4.32a, b)$$

It should be noted that in obtaining Eq. (4.31) one assumes that $\zeta_r < 1$. That is, the system is lightly damped. For critical damping of mode r, $\zeta_r = 1$. For the heavily damped case, $\zeta_r > 1$, Eq. (4.31) can be modified. But a mode in which $\zeta_r > 1$ will contribute a very insignificant amount to the response and therefore this case is not included here. The integral in Eq. (4.31) can be evaluated once the forcing function is provided.

Similar to the system in Section 4.5, the responses for the damped system are found by the following relation

$$\begin{pmatrix}x_1\\x_2\end{pmatrix} = [R]\begin{pmatrix}q_1\\q_2\end{pmatrix} = \frac{1}{\sqrt{2m}}\begin{bmatrix}1 & 1\\1 & -1\end{bmatrix}\begin{pmatrix}q_1\\q_2\end{pmatrix}. \qquad (4.33)$$

Now, returning to the determination of λ_m and λ_k for the present system, one applies Eq. $(4.32b)$,

$$2\zeta_r\omega_r = \lambda_m + \lambda_k\omega_r^2 .$$

Assume the specific damping parameters $\zeta_1 = \zeta_2 = 0.05$ which means that the modal damping parameter for each of these two modes is 5% critical. Therefore, upon substituting for ζ_1 and ω_1 into this equation, one has

$$0.10\left(2\sqrt{k/m}\,\right) = \lambda_m + \lambda_k\left(2\sqrt{k/m}\,\right)^2.$$

Similarly, upon substituting for ζ_2 and ω_2 into Eq. $(4.32b)$, one obtains

$$0.10\left(\sqrt{6k/m}\,\right) = \lambda_m + \lambda_k\left(\sqrt{6k/m}\,\right)^2.$$

Solving these two equations, one finds

$$\lambda_m = \left(\frac{0.20\,\sqrt{6}}{\sqrt{4}+\sqrt{6}}\right)\sqrt{\frac{k}{m}} = 0.1101\sqrt{\frac{k}{m}},$$
$$\lambda_k = \left(\frac{0.10}{\sqrt{4}+\sqrt{6}}\right)\sqrt{\frac{m}{k}} = 0.02247\sqrt{\frac{m}{k}}.$$

Remarks:

- In the foregoing tdof system, the modal damping parameters ζ_r are specified for two modes and the two constants λ_m and λ_k are found so that the proportional damping condition in Eq. (4.28) is satisfied. For systems with more than tdof and with the modal damping parameters specified for the first two modes, the damping parameters associated with modes higher than the first two are related to the damping parameters of the first two modes. In other words, if the higher modal damping parameters are to be chosen independently more constant terms are required in Eq. (4.28). For discussion on this case, the readers are referred Chapter 5 or to the references (Warburton, 1976, 64-66) (Clough and Penzien, 1975, 195).

- For linear mdof systems whose dampings are not proportional the analytical solution steps are relatively much more tedious. Two analytical approaches for the solution of such systems are available. The first approach of decoupling the systems of governing equations of motion is due to Ma and associates ((Ma and Imam, 2010, **329**: 3182-3202). It requires a considerable amount of algebraic manipulation and it is beyond the scope of the present book. The second is the state vector approach commonly adopted in the field of engineering control. In the following section this approach is illustrated for a tdof system.

4.7 Forced Vibration of Tdof Systems with Non-Proportional Damping

It may be appropriate to point out that the mechanical impedance method presented in Section 4.3 has also been applied to a tdof system with non-proportional damping. In this section the *space vector* or *state vector* approach for systems with non-proportional damping is presented.

Consider a simplified model of a rotating washing machine, which is a tdof system governed by the following matrix equation of motion as

$$[M](\ddot{x}) + [C](\dot{x}) + [K](x) = (f) \qquad (4.34)$$

where the mass, damping, stiffness matrices, and displacement as well as forcing vector are given, respectively by

$$[M] = \begin{bmatrix} m & 0 \\ 0 & m \end{bmatrix}, \qquad [C] = \begin{bmatrix} 0 & -2m\Omega \\ 2m\Omega & 0 \end{bmatrix},$$

$$[K] = \begin{bmatrix} k - m\Omega^2 & 0 \\ 0 & k - m\Omega^2 \end{bmatrix}, \qquad (x) = \begin{pmatrix} x_1 \\ x_2 \end{pmatrix}, \qquad (f) = \begin{pmatrix} f_1 \\ f_2 \end{pmatrix}$$

in which $f_1 = \tilde{u}(t)$ is the unit step function, and $f_2 = 0$. In this model the damping matrix $[C]$ is skew symmetric and is known as the gyroscopic matrix. Note that this damping matrix is not similar to that given by Eq. (4.28).

To provide a more concrete illustration of the steps involved in solving for the responses of the system, the system parameters are provided. Suppose the mass m =10 kg, k = 20,000 N/m, and the angular speed of the rotor $\Omega = 10$ rad/s. The system starts from rest. The responses are found by three steps. The first step is the transformation of the system of second order differential equations of motion into the state vector form in addition to the solution of the associated eigenvalues and eigenvectors. The second step is concerned with the process of diagonalization of the *coefficient* or *amplification matrix* of the system. The third and final step is evaluation of the responses.

Step 1: ***Solution of eigenvalues and eigenvectors***

Let $(Y) = \begin{pmatrix} y_1 \\ y_2 \\ y_3 \\ y_4 \end{pmatrix} = \begin{pmatrix} x_1 \\ x_2 \\ \dot{x}_1 \\ \dot{x}_2 \end{pmatrix}$ where $\begin{pmatrix} \dot{y}_3 \\ \dot{y}_4 \end{pmatrix} = \begin{pmatrix} \ddot{x}_1 \\ \ddot{x}_2 \end{pmatrix}$, $\begin{pmatrix} y_1 \\ y_2 \end{pmatrix} = \begin{pmatrix} x_1 \\ x_2 \end{pmatrix}$.

The two second order differential equations are transformed to four first order differential equations. Thus, Eq. (4.34) are written in state space form as

$$\begin{pmatrix} \dot{y}_1 \\ \dot{y}_2 \\ \dot{y}_3 \\ \dot{y}_4 \end{pmatrix} = \begin{bmatrix} [0] & [I] \\ -[M]^{-1}[K] & -[M]^{-1}[C] \end{bmatrix} \begin{pmatrix} y_1 \\ y_2 \\ y_3 \\ y_4 \end{pmatrix} + \begin{pmatrix} 0 \\ 0 \\ \tilde{u}(t) \\ 0 \end{pmatrix} \qquad (4.35a)$$

where the coefficient or amplification matrix is given by

$$[A] = \begin{bmatrix} [0] & [I] \\ -[M]^{-1}[K] & -[M]^{-1}[C] \end{bmatrix},$$

where $[I] = \begin{bmatrix} 1 & 0 \\ 0 & 1 \end{bmatrix}$ is the *unit* or *identity matrix*. The inverse of the mass matrix

$$[M]^{-1} = \begin{bmatrix} 10 & 0 \\ 0 & 10 \end{bmatrix}^{-1} = \begin{bmatrix} 0.1 & 0 \\ 0 & 0.1 \end{bmatrix}.$$

Therefore,

$$[M]^{-1}[K] = \begin{bmatrix} 0.1 & 0 \\ 0 & 0.1 \end{bmatrix} \begin{bmatrix} 20{,}000 - 1000 & 0 \\ 0 & 20{,}000 - 1000 \end{bmatrix}$$

$$= \begin{bmatrix} 1900 & 0 \\ 0 & 1900 \end{bmatrix}.$$

Similarly,

$$[M]^{-1}[C] = \begin{bmatrix} 0.1 & 0 \\ 0 & 0.1 \end{bmatrix} \begin{bmatrix} 0 & -200 \\ 200 & 0 \end{bmatrix} = \begin{bmatrix} 0 & -20 \\ 20 & 0 \end{bmatrix}.$$

By substituting all the above, one has the amplification matrix

$$[A] = \begin{bmatrix} 0 & 0 & 1 & 0 \\ 0 & 0 & 0 & 1 \\ -1900 & 0 & 0 & 20 \\ 0 & -1900 & -20 & 0 \end{bmatrix}. \qquad (4.35b)$$

The eigenvalues of $[A]$ are obtained from the frequency equation, $|s[I]_4 - [A]| = 0$ or $|[A] - s\,[I]_4| = 0$ in which $[I]_4$ is the unit matrix of order 4 and s is a constant parameter. Thus,

$$\begin{vmatrix} -s & 0 & 1 & 0 \\ 0 & -s & 0 & 1 \\ -1900 & 0 & -s & 20 \\ 0 & -1900 & -20 & -s \end{vmatrix} = 0. \qquad (4.36)$$

Operating on Eq. (4.36), one obtains

$$-s \begin{vmatrix} -s & 0 & 1 \\ 0 & -s & 20 \\ -1900 & -20 & -s \end{vmatrix} + (1) \begin{vmatrix} 0 & -s & 1 \\ -1900 & 0 & 20 \\ 0 & -1900 & -s \end{vmatrix} = 0$$

$$s^4 + 1900\,s^2 + 400\,s^2 + 1900^2 + 1900\,s^2 = 0, \qquad \text{or}$$

$$s^4 + 4200\,s^2 + 3610000 = 0.$$

This is a quadratic equation in s^2. Thus, if one writes $\lambda = s^2$ so that the last equation becomes

$$\lambda^2 + 4200\,\lambda + 3610000 = 0. \qquad (4.37)$$

Solving this quadratic equation and, in turn, taking the square roots of the solutions, the eigenvalues of the system are

$$s_1 = -10(2\sqrt{5} - 1)\,i = -34.7214\,i, \quad s_2 = 10(2\sqrt{5} - 1)\,i = 34.7214\,i,$$

$$s_3 = -10(2\sqrt{5} + 1)\,i = -54.7214\,i, \quad s_4 = 10(2\sqrt{5} + 1)\,i = 54.7214\,i.$$

Clearly, there are no real parts in all the four eigenvalues which are not repeated roots. Therefore, the system is stable.

For the eigenvectors or mode shapes, one makes use

$$([A]-s[I]_4)(Y)=(0) \qquad \text{or}$$

$$\begin{bmatrix} -s & 0 & 1 & 0 \\ 0 & -s & 0 & 1 \\ -1900 & 0 & -s & 20 \\ 0 & -1900 & -20 & -s \end{bmatrix}\begin{pmatrix} Y_1 \\ Y_2 \\ Y_3 \\ Y_4 \end{pmatrix}=\begin{pmatrix} 0 \\ 0 \\ 0 \\ 0 \end{pmatrix}, \tag{4.38}$$

where $Y_3=\dot{Y}_1$, $Y_4=\dot{Y}_2$. Substituting $s=s_1=-10(2\sqrt{5}-1)\,i$ in Eq. (4.38), one finds

$$\begin{pmatrix} Y_1 \\ Y_2 \\ Y_3 \\ Y_4 \end{pmatrix}=\begin{pmatrix} 1 \\ i \\ -34.7214\,i \\ 34.7214 \end{pmatrix} Y_1.$$

Assume $Y_1=1$ so this eigenvector can be written as

$$\begin{pmatrix} Y_1 \\ Y_2 \\ Y_3 \\ Y_4 \end{pmatrix}_{(1)}=\begin{pmatrix} 1 \\ i \\ -34.7214\,i \\ 34.7214 \end{pmatrix}. \tag{4.39a}$$

The subscript (1) on the lhs of Eq. (4.39a) denotes the eigenvector associated with s_1.

Similarly, the remaining eigenvectors can be obtained as

$$\begin{pmatrix} Y_1 \\ Y_2 \\ Y_3 \\ Y_4 \end{pmatrix}_{(2)}=\begin{pmatrix} 1 \\ -i \\ 34.7214\,i \\ 34.7214 \end{pmatrix}, \quad \begin{pmatrix} Y_1 \\ Y_2 \\ Y_3 \\ Y_4 \end{pmatrix}_{(3)}=\begin{pmatrix} 1 \\ -i \\ -54.7214\,i \\ -54.7214 \end{pmatrix}, \tag{4.39b, c}$$

$$\begin{pmatrix} Y_1 \\ Y_2 \\ Y_3 \\ Y_4 \end{pmatrix}_{(4)}=\begin{pmatrix} 1 \\ i \\ 54.7214\,i \\ -54.7214 \end{pmatrix}. \tag{4.39d}$$

Step 2: ***Diagonalization of amplification matrix***

In Eq. (4.35) the solution can be obtained by transforming the state vector $\begin{pmatrix} y_1 \\ y_2 \\ y_3 \\ y_4 \end{pmatrix}$ into another vector. Let the new vector be related to $\begin{pmatrix} y_1 \\ y_2 \\ y_3 \\ y_4 \end{pmatrix} = [R]\begin{pmatrix} z_1 \\ z_2 \\ z_3 \\ z_4 \end{pmatrix}$, where the transformation matrix $[R]$ whose columns are the eigenvectors in Eq. (4.39). That is,

$$[R] = \left[\begin{pmatrix} Y_1 \\ Y_2 \\ Y_3 \\ Y_4 \end{pmatrix}_{(1)} \begin{pmatrix} Y_1 \\ Y_2 \\ Y_3 \\ Y_4 \end{pmatrix}_{(2)} \begin{pmatrix} Y_1 \\ Y_2 \\ Y_3 \\ Y_4 \end{pmatrix}_{(3)} \begin{pmatrix} Y_1 \\ Y_2 \\ Y_3 \\ Y_4 \end{pmatrix}_{(4)} \right].$$

Therefore,

$$[R] = \begin{bmatrix} 1 & 1 & 1 & 1 \\ i & -i & -i & i \\ -34.7214\,i & 34.7214\,i & -54.7214\,i & 54.7214\,i \\ 34.7214 & 34.7214 & -54.7214 & -54.7214 \end{bmatrix}. \tag{4.40}$$

Substituting the transformation into Eq. (4.35a), one obtains

$$[R]\begin{pmatrix} \dot{z}_1 \\ \dot{z}_2 \\ \dot{z}_3 \\ \dot{z}_4 \end{pmatrix} = [A][R]\begin{pmatrix} z_1 \\ z_2 \\ z_3 \\ z_4 \end{pmatrix} + \begin{pmatrix} 0 \\ 0 \\ \tilde{u}(t) \\ 0 \end{pmatrix}. \tag{4.41}$$

Pre-multiplying the inverse of $[R]$, it becomes

$$\begin{pmatrix} \dot{z}_1 \\ \dot{z}_2 \\ \dot{z}_3 \\ \dot{z}_4 \end{pmatrix} = [R]^{-1}[A][R]\begin{pmatrix} z_1 \\ z_2 \\ z_3 \\ z_4 \end{pmatrix} + [R]^{-1}\begin{pmatrix} 0 \\ 0 \\ \tilde{u}(t) \\ 0 \end{pmatrix}. \tag{4.42}$$

It will be shown in the following that $[D] = [R]^{-1}[A][R]$ is a diagonal matrix such that Eq. (4.42) is uncoupled and therefore z_r, $r = 1, 2, 3, 4$ can be evaluated individually.

Now, $[R]^{-1}$ is required. It can be shown that

$$[R]^{-1} = \begin{bmatrix} 0.3059 & -0.3059\,i & 0.0056\,i & 0.0056 \\ 0.3059 & 0.3059\,i & -0.0056\,i & 0.0056 \\ 0.1941 & 0.1941\,i & 0.0056\,i & -0.0056 \\ 0.1941 & -0.1941\,i & -0.0056\,i & -0.0056 \end{bmatrix}. \qquad (4.43)$$

By making use of Eqs. (4.43), (4.35b), and (4.40) the diagonal matrix becomes

$$[D] = [R]^{-1}[A][R]$$

$$= \begin{bmatrix} -34.7214\,i & 0.0 & 0.0 & 0.0 \\ 0 & 34.72\,14i & 0.0 & 0 \\ 0 & 0.0 & -54.7214\,i & 0 \\ 0.0 & 0 & 0 & 54.7214\,i \end{bmatrix}. \qquad (4.44)$$

Note that the diagonal elements of this diagonal matrix are the eigenvalues of $[A]$.

Step 3: *Forced responses*
Returning to Eq. (4.42),

$$\dot{z}_1 = (-34.7214\,i)z_1 + (0.0056\,i)\tilde{u}(t).$$

$$\dot{z}_2 = (34.72\,14\,i)z_2 - (0.0056\,i)\tilde{u}(t).$$

$$\dot{z}_3 = (-54.7214\,i)z_3 + (0.0056\,i)\tilde{u}(t).$$

$$\dot{z}_4 = (54.7214\,i)z_4 - (0.0056\,i)\tilde{u}(t).$$

The first solution is

$$z_1 = c_1 e^{s_1 t} + (0.0056\,i)e^{s_1 t}\int_0^t e^{-s_1 \tau}\,\tilde{u}(\tau)\,d\tau, \qquad (4.45a)$$

where $s_1 = -34.7214\,i$, $\tilde{u}(\tau)$ is the unit step function, and c_1 is a constant to be found by applying the initial conditions. Therefore, Eq. (4.45a) becomes

$$z_1 = c_1 e^{-34.7214\,i\,t} + (0.0056\,i) e^{-34.7214\,i\,t} \int_0^t e^{\,34.7214\,i\,\tau}\, d\tau,$$

$$= c_1 e^{-34.7214\,i\,t} + \frac{(0.0056\,i) e^{-34.7214\,i\,t} e^{\,34.7214\,i\,t}}{34.7214\,i}$$

$$= c_1 e^{-34.7214\,i\,t} + 1.6128 \times 10^{-4}\,. \qquad (4.45b)$$

Similarly, one finds

$$z_2 = c_2 e^{34.7214\,i\,t} + 1.6128 \times 10^{-4}\,, \qquad (4.45c)$$

$$z_3 = c_3 e^{-54.7214\,i\,t} + 1.0234 \times 10^{-4}\,, \qquad (4.45d)$$

$$z_4 = c_4 e^{54.7214\,i\,t} + 1.0234 \times 10^{-4}. \qquad (4.45e)$$

Now, the responses for the tdof system are obtained by making use of Eq. (4.40),

$$\begin{pmatrix} y_1 \\ y_2 \\ y_3 \\ y_4 \end{pmatrix} = [R] \begin{pmatrix} z_1 \\ z_2 \\ z_3 \\ z_4 \end{pmatrix}$$

which gives, recall that $(y_1 \quad y_2 \quad y_3 \quad y_4)^T = (x_1 \quad x_2 \quad \dot{x}_1 \quad \dot{x}_2)^T$,

$$\begin{pmatrix} y_1 \\ y_2 \\ y_3 \\ y_4 \end{pmatrix} = \begin{pmatrix} z_1 + z_2 + z_3 + z_4 \\ z_1 i - z_2 i - z_3 i + z_4 i \\ -34.7214\,i\,z_1 + 34.7214\,i\,z_2 - 54.7214\,i\,z_3 + 54.7214\,i\,z_4 \\ 34.7214\,z_1 + 34.7214\,z_2 - 54.7214\,z_3 - 54.7214\,z_4 \end{pmatrix}.$$

After applying Eqs. (4.45*b*) through (4.45*e*), one has

$$y_1 = c_1 e^{-34.7214\,i\,t} + 1.6128 \times 10^{-4} + c_2 e^{34.7214\,i\,t} + 1.6128 \times 10^{-4}$$

$$+ c_3 e^{-54.7214\,i\,t} + 1.0234 \times 10^{-4} + c_4 e^{54.7214\,i\,t} + 1.0234 \times 10^{-4}$$

$$= c_1 e^{-34.7214\,i\,t} + c_2 e^{34.7214\,i\,t} + c_3 e^{-54.7214\,i\,t} + c_4 e^{54.7214\,i\,t}$$

$$+ 5.2724 \times 10^{-4}, \qquad (4.46a)$$

$$y_2 = i\,(c_1 e^{-34.7214\,i\,t} - c_2 e^{34.7214\,i\,t} - c_3 e^{-54.7214\,i\,t} + c_4 e^{54.7214\,i\,t}), \qquad (4.46b)$$

$$y_3 = i\,(-\,34.7214\,c_1 e^{-\,34.7214\,i\,t} + 34.7214\,c_2 e^{34.7214\,i\,t} \\ -\,54.7214\,c_3 e^{-\,54.7214\,i\,t} + 54.7214\,c_4 e^{54.7214\,i\,t}), \qquad (4.46c)$$

$$y_4 = 34.7214\,c_1 e^{-\,34.7214\,i\,t} + 34.7214\,c_2 e^{34.7214\,i\,t}$$

$$-\,54.7214\,c_3 e^{-\,54.7214\,i\,t} - 54.7214\,c_4 e^{54.7214\,i\,t}. \qquad (4.46d)$$

Since the system starts from rest, it means $\begin{pmatrix} x_1(0) \\ x_2(0) \\ \dot{x}_1(0) \\ \dot{x}_2(0) \end{pmatrix} = \begin{pmatrix} y_1(0) \\ y_2(0) \\ y_3(0) \\ y_4(0) \end{pmatrix} = \begin{pmatrix} 0 \\ 0 \\ 0 \\ 0 \end{pmatrix}$.

Applying these initial conditions to Eq. (4.46), one obtains

$$0 = c_1 + c_2 + c_3 + c_4 + 5.2724 \times 10^{-4}, \qquad (4.47a)$$

$$0 = i\,(c_1 - c_2 - c_3 + c_4), \qquad (4.47b)$$

$$0 = i\,(-\,34.7214\,c_1 + 34.7214\,c_2 - 54.7214\,c_3 + 54.7214\,c_4), \qquad (4.47c)$$

$$0 = 34.7214\,c_1 + 34.7214\,c_2 - 54.7214\,c_3 - 54.7214\,c_4. \qquad (4.47d)$$

Solving for the constants, one finds

$$c_1 = c_2 = -1.6128 \times 10^{-4}, \qquad c_3 = c_4 = -1.0234 \times 10^{-4}.$$

Substituting these constants into Eq. (4.46) and after some algebraic manipulation, one finds

$$x_1 = 10^{-4}\,[\,5.2724 - 3.2256\,cos(34.7214\,t) \\ -\,2.0468\,cos(54.7214\,t)], \qquad (4.48a)$$

$$x_2 = 10^{-4}\,[\,2.0468\,sin(54.7214\,t) \\ -\,3.2256\,sin(34.7214\,t)], \qquad (4.48b)$$

$$\dot{x}_1 = 1.1200 \times 10^{-2}\,[\,sin(34.7214\,t) \\ +\,sin(54.7214\,t)\,], \qquad (4.48c)$$

$$\dot{x}_2 = 1.1200 \times 10^{-2}\,[\,\cos(54.7214t) \\ -\,\cos(34.7214\,t)\,]. \qquad (4.48d)$$

As indicated in the foregoing, even for a tdof system the amount of algebraic manipulation by using the state vector approach is substantial and time consuming in evaluating the responses. Thus, for systems with many dof a numerical software package such as MatLab is required for the determination of the system responses.

Appendix 4A

Derivation of Natural Frequencies of a Tdof System

Consider first the determinant of the impedance matrix $[Z(\omega)]$,

$$|Z(\omega)| = \begin{vmatrix} (k_{11} - m_1\omega^2) & k_{12} \\ k_{21} & (k_{22} - m_2\omega^2) \end{vmatrix},$$

That is,

$$|Z(\omega)| = (k_{11} - m_1\omega^2)(k_{22} - m_2\omega^2) - k_{12}k_{21}.$$

Therefore, the chracterestic equation of the system is

$$k_{11}k_{22} - k_{11}m_2\omega^2 - m_1\omega^2k_{22} + m_1m_2\omega^4 - k_{12}k_{21} = 0.$$

Rearranging into a polynomial in ω, one has

$$(m_1m_2)\omega^4 - (k_{11}m_2 + k_{22}m_1)\omega^2 + (k_{11}k_{22} - k_{12}k_{21}) = 0.$$

Rewrite this equation as

$$a\omega^4 + b\omega^2 + c = 0 \qquad (A.1)$$

where $a = m_1m_2$, $b = -(k_{11}m_2 + k_{22}m_1)$, and $c = k_{11}k_{22} - k_{12}k_{21}$.

Solving for Eq. $(A.1)$, one obtains

$$\omega_1^2 = \frac{-b + \sqrt{b^2 - 4ac}}{2a}, \quad \omega_2^2 = \frac{-b - \sqrt{b^2 - 4ac}}{2a}. \qquad (A.2a, b)$$

The objective now is to show that Eq. (4.4b) is true. That is,

$$|Z(\omega)| = m_1m_2\,(\omega_1^2 - \omega^2)(\omega_2^2 - \omega^2).$$

To this end, one expands the rhs to give

$$|Z(\omega)| = (m_1m_2)\omega^4 + m_1m_2\,(-\omega_1^2 - \omega_2^2)\omega^2 + m_1m_2\,\omega_1^2\omega_2^2$$

Since ω_1 and ω_2 are the natural frequencies of the tdof system and therefore the last equation should agree with Eq. $(A.1)$. This means that one has to show that

$$b = m_1 m_2 \left(-\omega_1^2 - \omega_2^2\right), \qquad \text{and} \qquad c = m_1 m_2\, \omega_1^2 \omega_2^2. \qquad (A.3a, b)$$

Now, substituting Eq. (*A*.2) into the rhs of Eq. (*A*.3*a*), it results

$$-m_1 m_2 \left(\frac{-2b}{2a}\right) = m_1 m_2 \left(\frac{b}{a}\right).$$

However, from Eq. (*A*.1), $a = m_1 m_2$ and therefore the rhs of Eq. (*A*.3 a) = the lhs of Eq. (*A*.3 a).

Substituting Eq. (*A*.2) into the rhs of Eq. (*A*.3*b*),

$$m_1 m_2\, \omega_1^2 \omega_2^2 = m_1 m_2 \left(\frac{-b + \sqrt{b^2 - 4ac}}{2a}\right)\left(\frac{-b - \sqrt{b^2 - 4ac}}{2a}\right)$$

$$= m_1 m_2 \left(\frac{1}{4a}\right)\left(b^2 - b^2 + 4ac\right) = m_1 m_2 \left(\frac{c}{a}\right) = c.$$

Thus, the rhs of Eq. (*A*.3*b*) = lhs of Eq. (*A*.3*b*). This completes the proof of Eq. (4.4*b*).

References

Clough, R. W., and Penzien, J. 1975. Dynamics of Structures. New York: McGraw-Hill Book Co.

Ma, F., Morzfeld, M., and Imam, A. 2010. "The decoupling of damped linear systems in free and forced vibration". Journal of Sound and Vibration, **329**: 3182-3202.

Warburton, G. B. 1976. The Dynamical Behaviour of Structures, Second Edition. Oxford, England: Pergamon Press.

Exercise Questions

Q9. Applying Newton's second law of motion, derive the equation of motion for the system shown in Figure 4Q9. Determine the undamped natural frequencies of the system in term of coefficient of spring *k*, mass *m*, and length *L*.

If the stiffness of the spring system attached to the right-hand end of the rigid bar approaches to an infinitely large value and the remaining spring constant (that is, the stiffness of the dynamic absorber) does not change what is its implication on the two natural frequencies of the system?

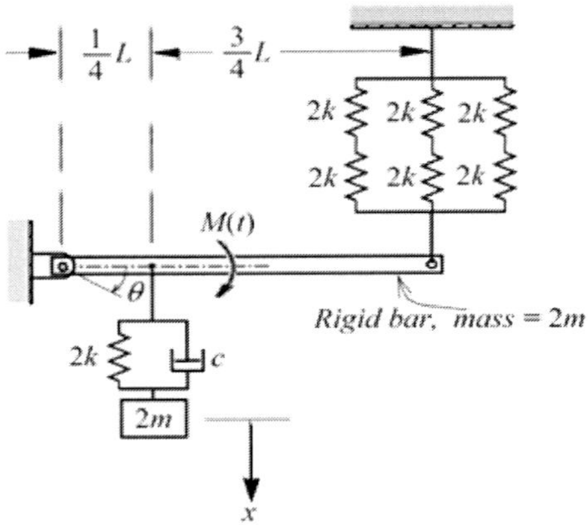

Figure 4Q9: A hinged rigid bar with a dynamic absorber.

Q10. Applying Newton's second of law of motion and related principle, derive the equations of motion, in the plane of the paper (suppose, the rods, masses, and connected springs are on a frictionless table), of the system shown in Figure 4Q10. It is assumed that the horizontal rods are rigid and their masses are zero, and the amplitudes of oscillation are small. Each of the two horizontal rods is hinged at one end and the other end is attached to a concentrated mass *m*.

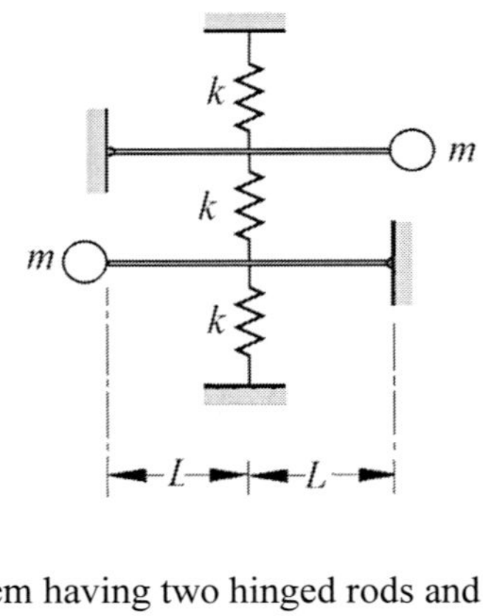

Figure 4Q10: A system having two hinged rods and connected to springs.

Q11. A double pendulum of lengths L_1 and L_2 with masses m_1 and m_2 are shown in Figure 4Q11. Applying Newton's second law of motion derive the equations of motion. Find the natural frequencies of the double pendulum.

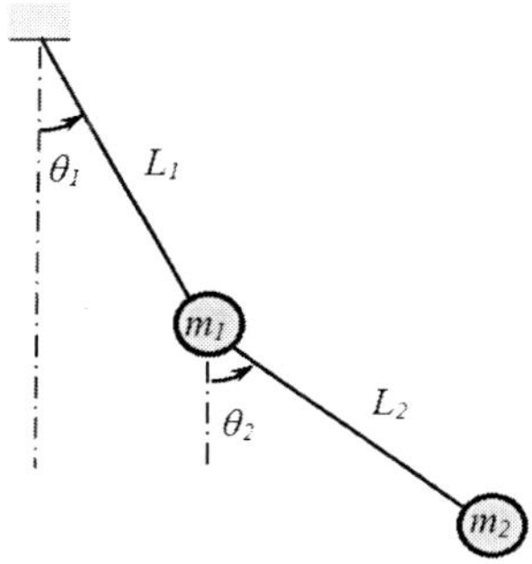

Figure 4Q11 : A double pendulum.

Q12. A single cylinder reciprocating engine of mass m_1 is mounted on a fixed-fixed beam of length L, width w, thickness h, and Young's modulus E, as shown in Figure 4Q12. A spring-mass system having mass m_2 and stiffness k_2 is attached to the beam at its center as indicated in the figure.

Assuming the beam has no mass and the steady-state harmonic force from the reciprocating engine is acted on at the mid-position of the beam.

(a) Derive the equations of motion for the above system.

(b) Find the relation between m_2 and k_2 that leads to no steady-state vibration of the reciprocating engine when a harmonic force from the engine is given by $f_1(t) = f_1 = F\cos\omega t$.

(c) Evaluate the steady-state displacement X_1 if $L = 10$ m, $w = 1$ m, $h = 2.10858$ cm, $E = 2 \times 10^{11}$ Pa, $k_2 = 10\ \frac{\text{kN}}{\text{m}}$, $m_1 = 1000$ kg, $m_2 = 100$ kg, $F = 100$ N, and $\omega = 5$ rad/s.

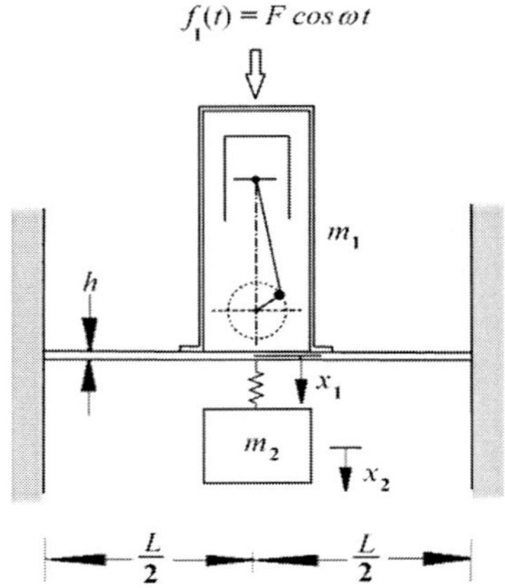

Figure 4Q12: A single cylinder reciprocating engine mounted on a fixed-fixed beam.

CHAPTER 5

FREE AND FORCED VIBRATION OF MANY DEGREES OF FREEDOM SYSTEMS

As seen in Chapter 4, the systems with two degrees of freedom (tdof) have been successfully considered as a refinement of the systems with single degree of freedom (sdof) for many situations. Naturally, we may have to consider a system with many degrees of freedom (mdof) when a sdof or tdof model is not adequate in general. In this chapter the focus is on the analysis of linear mdof systems. In Section 5.1, the Lagrange's equations are developed. The Lagrange's equations approach is employed. In Section 5.2, free vibration of mdof systems is presented by way of working details of three examples. Normal modes, orthogonality, and expansion theorem that are important to the solution of mdof systems are included in Section 5.3. Normal mode vibration and responses of undamped mdof systems are considered in Section 5.4 while responses of forced mdof systems with proportional damping are dealt with in Section 5.5. In Section 5.6, responses of forced mdof systems with non-proportional damping by mechanical impedance method are illustrated with a 3dof system. In Section 5.7, the formulation and steps in the space vector approach for responses of forced mdof systems with non-proportional damping are presented. The emphasis throughout this chapter is the detailed steps of the solution process.

5.1 Lagrange's Equations

In classical mechanics, there are two main reasons for working with Lagrange's equations instead of Newton's. The first is that Lagrange's equations apply to any coordinate system, whereas Newton's are confined to the inertial frame. The second is the ease of working with constraints in the Lagrangian system.

The Lagrangian formulation is based on the use of scalar functions, the kinetic energy T and potential energy U, and an infinitesimal expression, the virtual work associated with nonconservative forces. Owing to its use of generalized coordinates and generalized forces this formulation provides a broader point of view of mechanics and is usually referred to as *analytical mechanics*. The latter is hinged on the concept of *virtual displacements,* which led to the development of variational calculus. Thus, analytical mechanics is also regarded as the *variational approach to mechanics*.

Before the derivation of Lagrange's equations, it may be appropriate to mention that the latter equations are equivalent to Kane's equations (Amirouche, 1992, 187). In other words, the equations of motion for a system may be obtained by applying either Lagrange's equations or Kane's equations.

5.1.1 D'Alembert's principle

Consider Newton's second law of motion for a system of N particles. One can write

$$\vec{F}_i + \vec{f}_i - m_i\ddot{\vec{r}}_i = 0, \qquad i = 1, 2, \ldots, N \tag{5.1}$$

where $\vec{F}_i$ are the applied or external forces, $\vec{f}_i$ are the constraint or internal forces such as tensions in ropes, or normal forces applied by surfaces, and $-m_i\ddot{\vec{r}}_i$ are the inertia forces.

In order to derive Lagrange's equations, one can extend the virtual work principle to the dynamical case. Thus, applying Equation (5.1), one can write

$$\left(\vec{F}_i + \vec{f}_i - m_i\ddot{\vec{r}}_i\right) \bullet \delta\vec{r}_i = 0, \qquad i = 1, 2, \ldots, N$$

in which $\delta\vec{r}_i$ are the virtual displacements. Summing over the entire system,

$$\sum_{i=1}^{N}\left(\vec{F}_i + \vec{f}_i - m_i\ddot{\vec{r}}_i\right) \bullet \delta\vec{r}_i = 0, \tag{5.2}$$

Equation (5.2) is generally referred to as *d'Alembert's principle*. Assume that the virtual work of the constraint and internal forces vanishes, that is, $\sum_{i=1}^{N}\vec{f}_i \bullet \delta\vec{r}_i = 0$, then Equation (5.2) becomes

$$\sum_{i=1}^{N}\left(\vec{F}_i - m_i\ddot{\vec{r}}_i\right) \bullet \delta\vec{r}_i = 0, \tag{5.3}$$

Equation (5.3) is known as the *generalized principle of d'Alembert* (Meirovitch, 1997, 81). The term $\vec{F}_i - m_i\ddot{\vec{r}}_i$ is often called the *effective force*.

5.1.2 Lagrange's equations for conservative and nonconservative systems

It is understood that the position vectors of the particles are not always independent and therefore it is desirable to express their motions in terms of independent generalized coordinates. To begin with, let the position vectors $\vec{r}_i$ be explicitly independent of time so that for a system of n degrees of freedom ($n$dof) the coordinate transformation becomes

$$\vec{r}_i = \vec{r}_i(q_1, q_2, \ldots, q_n) \qquad i = 1, 2, \ldots, N. \tag{5.4}$$

The velocities $\dot{\vec{r}}_i$ are obtained by differentiating Equation (5.4) with respect to time,

$$\dot{\vec{r}}_i = \frac{d\vec{r}_i}{dt} = \frac{\partial\vec{r}_i}{\partial q_1}\left(\frac{dq_1}{dt}\right) + \frac{\partial\vec{r}_i}{\partial q_2}\left(\frac{dq_2}{dt}\right) + \cdots + \frac{\partial\vec{r}_i}{\partial q_n}\left(\frac{dq_n}{dt}\right)$$

$$= \sum_{k=1}^{n} \frac{\partial\vec{r}_i}{\partial q_k}\,\dot{q}_k \tag{5.5}$$

or $\dot{\vec{r}}_i = \sum_{k=1}^{n} \frac{\partial\vec{r}_i}{\partial q_k}\,\dot{q}_k$. According to Equation (5.4), $\vec{r}_i$ does not depend explicitly on generalized velocities $\dot{q}_k$, therefore partial differentiating the Equation (5.5) with respect to $\dot{q}_k$ yields

$$\frac{\partial(\dot{\vec{r}}_i)}{\partial\dot{q}_k} = \frac{\partial\vec{r}_i}{\partial q_k} \qquad i = 1, 2, \ldots, N;\ k = 1, 2, \ldots, n. \tag{5.6}$$

Furthermore, by analogy with Equation (5.5), one has

$$\delta\vec{r}_i = \frac{\partial\vec{r}_i}{\partial q_1}\,\delta q_1 + \frac{\partial\vec{r}_i}{\partial q_2}\,\delta q_2 + \cdots + \frac{\partial\vec{r}_i}{\partial q_n}\,\delta q_n = \sum_{k=1}^{n} \frac{\partial\vec{r}_i}{\partial q_k}\,\delta q_k. \tag{5.7}$$

By making use of Equation (5.7), the second term in Equation (5.3) can be written as

$$\sum_{i=1}^{N}(m_i\ddot{\vec{r}}_i)\bullet\delta\vec{r}_i=\sum_{i=1}^{N}(m_i\ddot{\vec{r}}_i)\bullet\sum_{k=1}^{n}\frac{\partial\vec{r}_i}{\partial q_k}\delta q_k$$
$$=\sum_{k=1}^{n}\left(\sum_{i=1}^{N}m_i\ddot{\vec{r}}_i\bullet\frac{\partial\vec{r}_i}{\partial q_k}\right)\delta q_k. \quad (5.8)$$

However, the differential operation,

$$\frac{d}{dt}\left(m_i\dot{\vec{r}}_i\bullet\frac{\partial\vec{r}_i}{\partial q_k}\right)=\frac{d}{dt}(m_i\dot{\vec{r}}_i)\bullet\frac{\partial\vec{r}_i}{\partial q_k}+(m_i\dot{\vec{r}}_i)\bullet\frac{d}{dt}\left(\frac{\partial\vec{r}_i}{\partial q_k}\right)$$
$$=m_i\ddot{\vec{r}}_i\bullet\frac{\partial\vec{r}_i}{\partial q_k}+(m_i\dot{\vec{r}}_i)\bullet\frac{d}{dt}\left(\frac{\partial\vec{r}_i}{\partial q_k}\right). \quad (5.9)$$

The first term on the rhs of Equation (5.9) is a typical term inside the brackets on the rhs of Equation (5.8). Thus, by moving the second term on the rhs of Equation (5.9) to its lhs, one has

$$m_i\ddot{\vec{r}}_i\bullet\frac{\partial\vec{r}_i}{\partial q_k}=\frac{d}{dt}\left(m_i\dot{\vec{r}}_i\bullet\frac{\partial\vec{r}_i}{\partial q_k}\right)-(m_i\dot{\vec{r}}_i)\bullet\frac{d}{dt}\left(\frac{\partial\vec{r}_i}{\partial q_k}\right). \quad (5.10)$$

Considering Equation (5.6) and the fact that the order of total derivatives with respect to time and partial derivatives with respect to q_k is interchangeable, one can write Equation (5.10) as

$$m_i\ddot{\vec{r}}_i\bullet\frac{\partial\vec{r}_i}{\partial q_k}=\frac{d}{dt}\left(m_i\dot{\vec{r}}_i\bullet\frac{\partial\dot{\vec{r}}_i}{\partial\dot{q}_k}\right)-(m_i\dot{\vec{r}}_i)\bullet\left(\frac{\partial\dot{\vec{r}}_i}{\partial q_k}\right)$$
$$=\frac{d}{dt}\left[m_i\frac{\partial}{\partial\dot{q}_k}\left(\frac{1}{2}\dot{\vec{r}}_i\bullet\dot{\vec{r}}_i\right)\right]-(m_i)\frac{\partial}{\partial q_k}\left(\frac{1}{2}\dot{\vec{r}}_i\bullet\dot{\vec{r}}_i\right)$$
$$=\left[\frac{d}{dt}\left(\frac{\partial}{\partial\dot{q}_k}\right)-\frac{\partial}{\partial q_k}\right]\left(\frac{1}{2}m_i\dot{\vec{r}}_i\bullet\dot{\vec{r}}_i\right), \quad (5.11)$$

where $\frac{1}{2}m_i\dot{\vec{r}}_i\bullet\dot{\vec{r}}_i$ is the kinetic energy of particle i. Therefore, substituting Equation (5.11) into (5.8) results in

$$\sum_{i=1}^{N}(m_i\ddot{\vec{r}}_i)\bullet\delta\vec{r}_i=\sum_{k=1}^{n}\left\{\left[\frac{d}{dt}\left(\frac{\partial}{\partial\dot{q}_k}\right)-\frac{\partial}{\partial q_k}\right]\left(\sum_{i=1}^{N}\frac{1}{2}m_i\dot{\vec{r}}_i\bullet\dot{\vec{r}}_i\right)\right\}\delta q_k$$
$$=\sum_{k=1}^{n}\left[\frac{d}{dt}\left(\frac{\partial T}{\partial\dot{q}_k}\right)-\frac{\partial T}{\partial q_k}\right]\delta q_k \quad (5.12)$$

where the kinetic energy of the entire system is

$$T = \sum_{i=1}^{N} \frac{1}{2} m_i \dot{\vec{r}}_i \bullet \dot{\vec{r}}_i = T(q_1, q_2, \dots, q_n, \dot{q}_1, \dot{q}_2, \dots, \dot{q}_2) \tag{5.13}$$

in which the transformation (5.5) has been referenced to.

Now, returning to the first term on the lhs of the virtual work expression in Equation (5.3). That is,

$$\sum_{i=1}^{N} \vec{F}_i \bullet \delta\vec{r}_i = \delta W. \tag{5.14}$$

By making use of Equation (5.7), this equation becomes

$$\sum_{i=1}^{N} \vec{F}_i \bullet \sum_{k=1}^{n} \frac{\partial \vec{r}_i}{\partial q_k} \delta q_k = \sum_{k=1}^{n} \left(\sum_{i=1}^{N} \vec{F}_i \bullet \frac{\partial \vec{r}_i}{\partial q_k} \right) \delta q_k = \delta W. \tag{5.15}$$

Interpreting the term inside the brackets in Equation (5.15) as the *generalized* forces $Q_k = \sum_{i=1}^{N} \vec{F}_i \bullet \frac{\partial \vec{r}_i}{\partial q_k}$ such that Equation (5.14) can be written as

$$\delta W = \sum_{k=1}^{n} Q_k \, \delta q_k. \tag{5.16}$$

It should be noted the adjective *generalized* applied in the present context is to indicate that the generalized forces can include moments or other quantities so that $Q_k \, \delta q_k$ has the units of work.

Now, consider the forces acting on the system can be grouped into conservative forces, which may be derived from the potential energy $U = U(q_1, q_2, \dots, q_n)$, and nonconservative forces, which cannot be obtained from the potential energy, then the first term on the lhs of Equation (5.3) reduces to

$$\sum_{i=1}^{N} \vec{F}_i \bullet \delta\vec{r}_i = \delta W = \delta W_c + \delta W_{nc} = -\,\delta U + \sum_{k=1}^{n} Q_{nck} \, \delta q_k, \tag{5.17}$$

where $\delta W_c = -\,\delta U = -\left(\frac{\partial U}{\partial q_1} \delta q_1 + \frac{\partial U}{\partial q_2} \delta q_2 + \cdots + \frac{\partial U}{\partial q_n} \right) \delta q_n$, and Q_{nck} for $k = 1,2, \dots, n$ are the nonconservative generalized forces.

Equation (5.17) can be written as

$$\sum_{i=1}^{N} \vec{F}_i \bullet \delta\vec{r}_i = -\sum_{k=1}^{n} \left(\frac{\partial U}{\partial q_k} - Q_{nck}\right) \delta q_k. \qquad (5.18)$$

Substituting Equations (5.12) and (5.18) into (5.3), one has

$$-\sum_{k=1}^{n} \left[\frac{d}{dt}\left(\frac{\partial T}{\partial \dot{q}_k}\right) - \frac{\partial T}{\partial q_k} + \frac{\partial U}{\partial q_k} - Q_{nck}\right] \delta q_k = 0 \qquad (5.19)$$

in which the generalized coordinates are independent from one another, the virtual displacements δq_k may assume arbitrary values. Therefore, Equation (5.19) will be satisfied if and only if the coefficient becomes zero. That is,

$$\frac{d}{dt}\left(\frac{\partial T}{\partial \dot{q}_k}\right) - \frac{\partial T}{\partial q_k} + \frac{\partial U}{\partial q_k} = Q_{nck} \qquad k = 1, 2, \ldots, n. \qquad (5.20)$$

Equation (5.20) is the Lagrange's equations of motion.

In general, the potential energy U does not depend upon the generalized velocities $\dot{q}_k$ and therefore, one can introduce the *Lagrangian* defined by $L = T - U$ such that Equation (5.20) may be cast into a more compact form as

$$\frac{d}{dt}\left(\frac{\partial L}{\partial \dot{q}_k}\right) - \frac{\partial L}{\partial q_k} = Q_{nck} \qquad k = 1, 2, \ldots, n. \qquad (5.21)$$

In the special case, some of the nonconservative generalized forces are those associated with the viscous damping. If the damping forces are proportional to the generalized velocities, it is convenient to use the so-called *Rayleigh's dissipation function* (Meirovitch, 1997, 390) which can be expressed as

$$D = \frac{1}{2}\sum_{r=1}^{n}\sum_{s=1}^{n} c_{rs}\,\dot{q}_r\dot{q}_s \qquad (5.22)$$

in which the coefficients c_{rs} are symmetric in r and s. This dissipation function enables one to derive viscous damping forces similar to that for conservative forces. For example, the viscous damping forces Q_{vk} may be derived from Rayleigh's dissipation function as

$$Q_{vk} = -\frac{\partial D}{\partial \dot{q}_k} \qquad k = 1, 2, \dots, n. \tag{5.23}$$

Suppose the nonconservative generalized forces in Equation (5.21) can be divided into dissipative forces and forces impressed on the system by external means, one can rewrite Equation (5.21) as

$$\frac{d}{dt}\left(\frac{\partial L}{\partial \dot{q}_k}\right) - \frac{\partial L}{\partial q_k} + \frac{\partial D}{\partial \dot{q}_k} = Q_{nk} \qquad k = 1, 2, \dots, n. \tag{5.24}$$

where now Q_{nk} is the generalized impressed forces.

In many situations, the system has no nonconservative forces such that $\frac{\partial D}{\partial \dot{q}_k} = 0$ and $Q_{nk} = 0$ for $k = 1, 2, \dots, n,$ then Lagrange's equations for conservative systems simply reduce to

$$\frac{d}{dt}\left(\frac{\partial L}{\partial \dot{q}_k}\right) - \frac{\partial L}{\partial q_k} = 0 \qquad k = 1, 2, \dots, n. \tag{5.25}$$

Example 5.1
Applying Lagrange's equations and assuming small oscillations, derive the equations of motion for the system shown in Figure 5.1.

Solution:
The kinetic energy of the system is

$$T = \frac{1}{2}m\,\dot{q}_1^2 + \frac{1}{2}m\,\dot{q}_2^2 + \frac{1}{2}\,J\dot{q}_3^2.$$

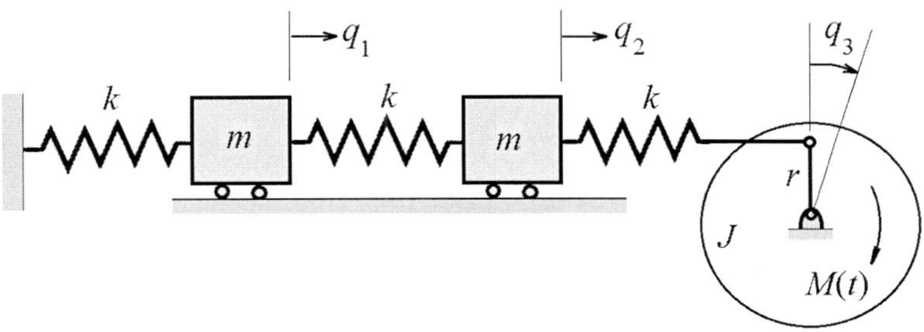

Figure 5.1: Mechanical system with 3dof.

The potential energy of the system is

$$U = \frac{1}{2}k\, q_1^2 + \frac{1}{2}k(q_2 - q_1)^2 + \frac{1}{2}k(rq_3 - q_2)^2.$$

The work done by the external moment, the generalized force

$$\delta W = M(t)\delta q_3 \quad \rightarrow \quad Q_{n3} = M(t).$$

The Lagrangian is given by

$$L = T - U = \frac{1}{2}m\, \dot{q}_1^2 + \frac{1}{2}m\, \dot{q}_2^2 + \frac{1}{2}\, J\dot{q}_3^2$$

$$-\frac{1}{2}k\, q_1^2 - \frac{1}{2}k(q_2 - q_1)^2 - \frac{1}{2}k(rq_3 - q_2)^2.$$

Applying Equation (5.24) and considering every term individually in the equations,

$$Q_{n1} = \; Q_{n2} = 0, \quad Q_{n3} = M(t), \quad \frac{\partial D}{\partial \dot{q}_k} = 0,$$

$$\frac{d}{dt}\left(\frac{\partial L}{\partial \dot{q}_1}\right) = m\ddot{q}_1, \quad \frac{d}{dt}\left(\frac{\partial L}{\partial \dot{q}_2}\right) = m\ddot{q}_2, \quad \frac{d}{dt}\left(\frac{\partial L}{\partial \dot{q}_3}\right) = J\ddot{q}_3,$$

$$\frac{\partial L}{\partial q_1} = -kq_1 - k(q_2 - q_1)(-1),$$

$$\frac{\partial L}{\partial q_2} = -k(q_2 - q_1)(1) - k(rq_3 - q_2)(-1), \frac{\partial L}{\partial q_3} = -k(rq_3 - q_2)(r),$$

Therefore, Equation (5.24), one has

$$m\ddot{q}_1 + kq_1 - k(q_2 - q_1) = 0,$$

$$m\ddot{q}_2 + k(q_2 - q_1) - k(rq_3 - q_2) = 0,$$

$$J\ddot{q}_3 + k(rq_3 - q_2)(r) = M(t).$$

Rearranging terms and expressing in matrix form, one obtains

$$\begin{bmatrix} m & 0 & 0 \\ 0 & m & 0 \\ 0 & 0 & J \end{bmatrix} \begin{pmatrix} \ddot{q}_1 \\ \ddot{q}_2 \\ \ddot{q}_3 \end{pmatrix} + \begin{bmatrix} 2k & -k & 0 \\ -k & 2k & -rk \\ 0 & -rk & r^2k \end{bmatrix} \begin{pmatrix} q_1 \\ q_2 \\ q_3 \end{pmatrix} = \begin{pmatrix} 0 \\ 0 \\ M(t) \end{pmatrix}.$$

With reference to this equation, the system has static coupling only.

5.1.3 Holonomic and nonholonomic constraints equations

In general, motions of bodies or particles are constrained in certain manners. In Newtonian dynamics, constraint forces are considered. Examples of the latter are tensions in cable or strings, and normal forces applied by surfaces. However, in Lagrangian dynamics the constraint forces can be easily incorporated in the formulation. But these constraints can be disregarded in the Lagrangian formulation. The constraints can be classified as *holonomic* and *nonholonomic*. Holonomic constraints are those concerned with geometric relations and nonholonomic ones are those dealing with inequalities conditions and velocity relations that cannot be solved by integrating them. In the following the Lagrange's equation approach for systems with holonomic and nonholonomic constraints are included to demonstrate its use of such classes of dynamic problems.

5.1.3.1 *Systems with holonomic constraints*

Holonomic constraints may be expressed in generalized coordinates as

$$f_j(q_1, q_2, \dots, q_n, t) = 0 \qquad j = 1, 2, \dots, 3N - n \tag{5.26}$$

where $3N$ is due to the fact that there are N particles in the 3D space having n dof.

In what follows the Lagrangian formulation is shown to deal with this type of constraints. To this end, $3N - n$ new variables λ_j which are called *Lagrange multipliers* and a new Lagrangian $L^{(1)}$ are introduced,

$$L^{(1)} = L + \lambda_j f_j \tag{5.27}$$

in which λ_j are treated as new coordinates. As L does not depend on λ_j, by applying Equation (5.27), becomes

$$\frac{\partial L^{(1)}}{\partial \lambda_j} = f_j = 0 \tag{5.28}$$

which is the constraint equations according to Equation (5.26). For a conservative system, application of Equation (5.27) in Equation (5.25) gives

$$\frac{d}{dt}\left(\frac{\partial L}{\partial \dot{q}_k}\right) - \frac{\partial L}{\partial q_k} = \sum_{j=1}^{3N-n} \lambda_j \frac{\partial f_j}{\partial q_k} \qquad k = 1, 2, \dots, n.$$

The rhs term can be written as $\sum_{j=1}^{3N-n} \lambda_j a_{jk}$ so that the above equation becomes

$$\frac{d}{dt}\left(\frac{\partial L}{\partial \dot{q}_k}\right) - \frac{\partial L}{\partial q_k} = \sum_{j=1}^{3N-n} \lambda_j a_{jk} \qquad k = 1, 2, \dots, n. \tag{5.29}$$

To provide a simple illustration, consider the pendulum moving in the plane as shown in Figure 5.2. The kinetic energy is $T = \frac{1}{2}\, m\, [(\dot{x})^2 + (\dot{y})^2]$, the potential energy $U = mgy$, and the holonomic constraint $f_1 = x^2 + y^2 - l^2 = 0$ such that the Lagrange multiplier is λ_1 or written simply as λ. Note, in this problem there is one particle and two dof. Therefore, $3N - n = 3(1) - 2 = 1$. Thus, $j = 1$.

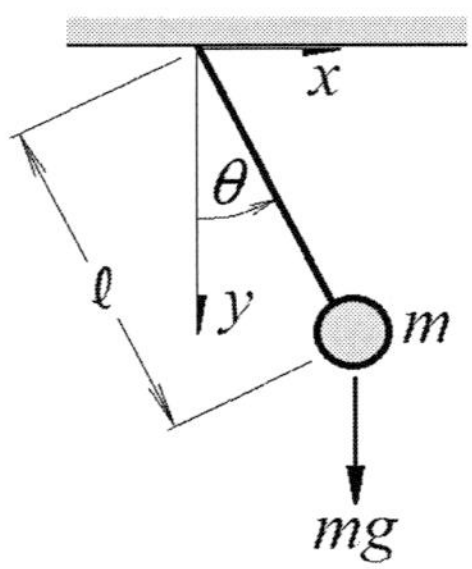

Figure 5.2: Pendulum moving in the plane.

The Lagrangian of the system is

$$L = \frac{1}{2}\, m\, [(\dot{x})^2 + (\dot{y})^2] + mgy. \tag{5.30}$$

Applying Equation (5.29) for this system, one obtains

$$m\ddot{x} = 2\lambda x, \quad m\ddot{y} = 2\lambda y + mg. \tag{5.31a, b}$$

Since there are only two differential equations and three variables (x, y, λ), one more differential equation is required in order to solve for these three variables.

By making use of the constraint equation and taking differentiation with respect to time twice, one has

$$x\,\ddot{x} + y\,\ddot{y} + (\dot{x})^2 + (\dot{y})^2 = 0. \tag{5.32}$$

Equations (5.31) and (5.32) are a complete set of three second order differential equations. One can therefore solve for λ to give

$$\lambda = -\frac{m}{2l^2}\left[gy + (\dot{x})^2 + (\dot{y})^2\right]. \tag{5.33}$$

Substituting this value into Equations $(5.31a, b)$, one can solve for x and y. Clearly, these two resulting second order differential equations are highly nonlinear and therefore, they can only be solved numerically.

Of course, the foregoing problem can be dealt with by using other approaches. For example, one can choose θ as a single generalized coordinate with no constraint. This is left to the reader and will not be further considered here.

5.1.3.2 *Systems with nonholonomic constraints*

A system to illustrate the nonholonomic constraints is the circular coin or disk of radius a rolling upright on the horizontal plane without slipping, as shown in Figure 5.3.

The nonholonomic constraints are due to the condition of without slipping. In this problem one can start by disregarding the nonholonomic constraints. The motion of the system may be described by four independent coordinates, x, y cartesian coordinates representing the position of the contact point $P(x, y)$ of the coin with the horizontal plane, ψ the angle between the axis of the coin and the x axis and θ the angle between a fixed radius of the coin and the vertical direction parallel to the plane of the coin. The Lagrangian of the system is therefore

$$L = \frac{1}{2}m\,\dot{x}^2 + \frac{1}{2}m\,\dot{y}^2 + \frac{1}{2}I_o\dot{\theta}^2 + \frac{1}{2}I_v\dot{\psi}^2 \tag{5.34}$$

where m is the mass of the coin, $I_o = ma^2$ and $I_v = \frac{1}{2}ma^2$ its moments of inertia about its own axis and a perpendicular direction, respectively.

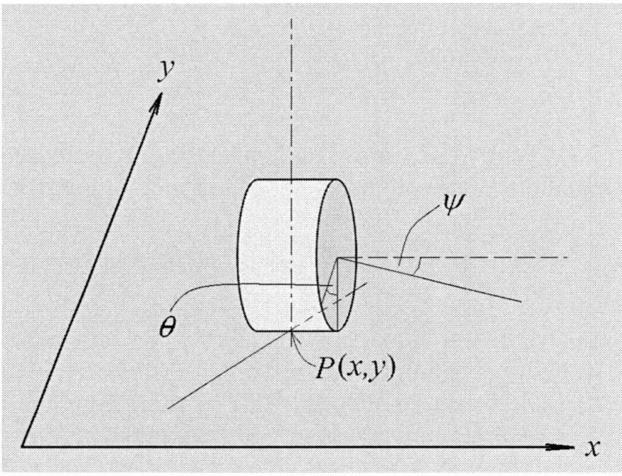

Figure 5.3: A coin rolls down an inclined plane.

The nonholonomic constraint can be specified via the equation of the velocity at the contact point,

$$v = a\dot{\theta}$$

which can be recast into the velocity components along the x and y directions as

$$\dot{x} = a\dot{\theta}\cos\psi, \qquad \dot{y} = a\dot{\theta}\sin\psi. \qquad (5.35a, b)$$

Since these two equations cannot be integrated to give constraints of the form $f_j = f_j(x, y, \psi, \theta) = 0$. Thus, they are nonholonomic.

Applying Equation (5.29), one can obtain the following equations

$$m\ddot{x} = \lambda_1, \quad m\ddot{y} = \lambda_2, \qquad (5.36a, b)$$

$$I_o\ddot{\theta} = -\lambda_1 a\sin\psi + \lambda_2 a\cos\psi, \qquad I_v\ddot{\psi} = 0. \qquad (5.36c, d)$$

These four equations contain six variables $(x, y, \psi, \theta, \lambda_1, \lambda_2)$ and therefore, two more independent equations are required for their solution. Equations $(5.35a, b)$ can be employed for this purpose. While there are six variables the system has only tdof. The two dof are the x and y. For brevity, the solution of these six variables will not be pursued here.

5.2 Free Vibration of Many Dof (Mdof) Systems and Equations of Motion

Consider the matrix equation of motion for free vibration of a *n*dof undamped system

$$[M](\ddot{\vec{x}}) + [K](\vec{x}) = (\vec{0}) \tag{5.37}$$

in which systems of tdof were dealt with in Section 4.1. In this section, free vibration of linear systems with *n*dof is studied by way of solving several problems in order to provide a more concrete understanding and detailed steps in the free vibration solution of such a system.

Example 5.2
Evaluate the natural frequencies of the 3dof spring-mass system as shown in Figure 5.4 in which $m_1 = 4m$, $m_2 = 2m$, $m_3 = m$, $k_1 = 3k$, $k_2 = k$, $k_3 = k$.

Solution:
The equations of motion can be obtained by applying Newton's second law of motion, $\sum F = ma$ to every mass. By making use of the FBD as shown

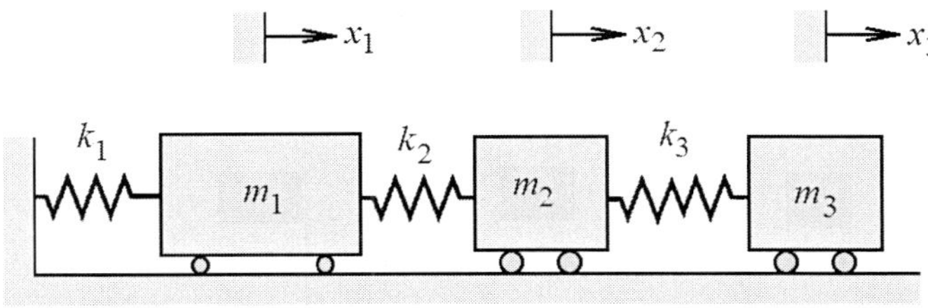

Figure 5.4: 3dof spring-mass system.

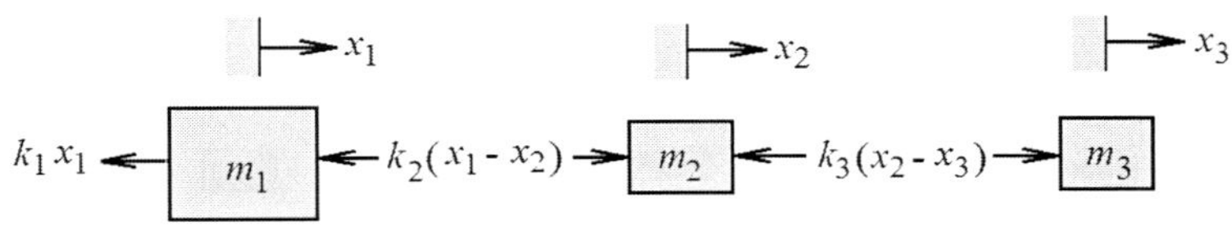

Figure 5.4a: Free-body diagram of spring-mass system in Figure 5.4.

in Figure 5.4a, one has

$$m_1\ddot{x}_1 = -k_1x_1 - k_2(x_1 - x_2),$$

$$m_2\ddot{x}_2 = k_2(x_1 - x_2) - k_3(x_2 - x_3),$$

$$m_3\ddot{x}_3 = k_3(x_2 - x_3).$$

Rearranging terms,

$$m_1\ddot{x}_1 + (k_1 + k_2)x_1 - k_2x_2 = 0,$$

$$m_2\ddot{x}_2 + (k_2 + k_3)x_2 - k_2x_1 - k_3x_3 = 0,$$

$$m_3\ddot{x}_3 - k_3x_2 + k_3x_3 = 0.$$

After substituting for the given data, it leads to

$$4m\ddot{x}_1 + 4kx_1 - kx_2 = 0, \qquad (i)$$

$$2m\ddot{x}_2 + 2kx_2 - kx_1 - kx_3 = 0, \qquad (ii)$$

$$m\ddot{x}_3 - kx_2 + kx_3 = 0. \qquad (iii)$$

Since the system has no damping and for oscillatory motion one can assume

$$x_1 = X_1 \cos \omega t, \quad x_2 = X_2 \cos \omega t, \quad x_3 = X_3 \cos \omega t$$

so that

$$\ddot{x}_1 = -\omega^2 X_1 \cos \omega t, \quad \ddot{x}_2 = -\omega^2 X_2 \cos \omega t, \quad \ddot{x}_3 = -\omega^2 X_3 \cos \omega t.$$

Substituting these values into Eqs. (i) through (iii), it results in

$$(4k - 4m\omega^2)X_1 - kX_2 = 0, \qquad (iv)$$

$$-kX_1 + (2k - 2m\omega^2)X_2 - kX_3 = 0, \qquad (v)$$

$$-kX_2 + (k - m\omega^2)X_3 = 0. \qquad (vi)$$

Writing in matrix form, one has

$$\begin{bmatrix} (4k - 4m\omega^2) & -k & 0 \\ -k & (2k - 2m\omega^2) & -k \\ 0 & -k & (k - m\omega^2) \end{bmatrix} \begin{pmatrix} X_1 \\ X_2 \\ X_3 \end{pmatrix} = \begin{pmatrix} 0 \\ 0 \\ 0 \end{pmatrix}. \quad (vii)$$

The frequency equation or characteristic equation is obtained by setting the determinant of the coefficient matrix to zero. That is,

$$\begin{vmatrix} (4k - 4m\omega^2) & -k & 0 \\ -k & (2k - 2m\omega^2) & -k \\ 0 & -k & (k - m\omega^2) \end{vmatrix} = 0$$

which, upon operation, becomes

$$8(k - m\omega^2)^3 - 5k^2(k - m\omega^2) = 0$$

$$(k - m\omega^2)[8(k - m\omega^2)^2 - 5k^2] = 0$$

$$(k - m\omega^2)[8(k^2 + m^2\omega^4 - 2km\omega^2) - 5k^2] = 0$$

$$(k - m\omega^2)(8m^2\omega^4 + 3k^2 - 16km\omega^2) = 0$$

which gives

$$\omega_1^2 = \frac{k}{m}, \quad \omega_2^2 = (1 - 0.7906)\frac{k}{m}, \quad \omega_3^2 = (1.7906)\frac{k}{m}.$$

Therefore, the natural frequencies of the system are

$$\omega_1 = \sqrt{\frac{k}{m}}, \quad \omega_2 = 0.4576\sqrt{\frac{k}{m}}, \quad \omega_3 = 1.3381\sqrt{\frac{k}{m}}.$$

Example 5.3

Evaluate the natural frequencies of the 3dof spring-mass system as shown in Figure 5.5 in which $m_1 = m_2 = m_3 = m$, $k_1 = k_2 = k_3 = k_4 = k_5 = k_6 = k$.

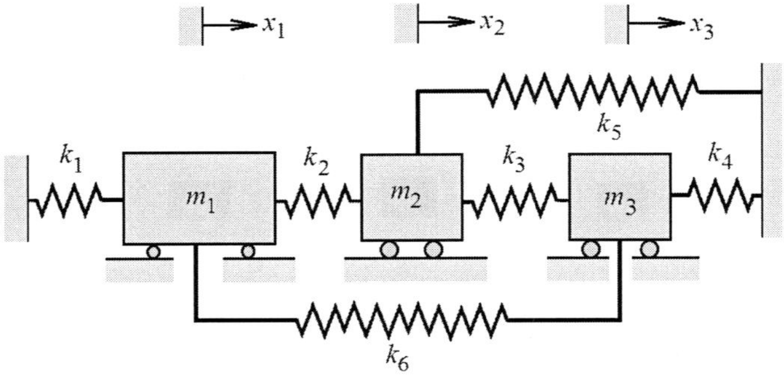

Figure 5.5: Spring-mass system with 3dof.

Solution:

The equations of motion can be obtained by applying Newton's second law of motion, $\sum F = ma$ to every mass. By making use of the FBD as shown in Figure 5.5a, one has

$$m_1\ddot{x}_1 = -k_1x_1 - k_2(x_1 - x_2) - k_6(x_1 - x_3),$$

$$m_2\ddot{x}_2 = -k_5x_2 + k_2(x_1 - x_2) - k_3(x_2 - x_3),$$

$$m_3\ddot{x}_3 = -k_4x_3 + k_3(x_2 - x_3) + k_6(x_1 - x_3).$$

Rearranging terms,

$$m_1\ddot{x}_1 + (k_1 + k_2 + k_6)x_1 - k_2x_2 - k_6x_3 = 0,$$

$$m_2\ddot{x}_2 + (k_2 + k_3 + k_5)x_2 - k_2x_1 - k_3x_3 = 0,$$

$$m_3\ddot{x}_3 - k_6x_1 - k_3x_2 + (k_3 + k_4 + k_6)x_3 = 0.$$

After substituting for the given data, it leads to

$$m\ddot{x}_1 + 3kx_1 - kx_2 - kx_3 = 0, \qquad (i)$$

$$m\ddot{x}_2 + 3kx_2 - kx_1 - kx_3 = 0, \qquad (ii)$$

$$m\ddot{x}_3 - kx_1 - kx_2 + 3kx_3 = 0. \qquad (iii)$$

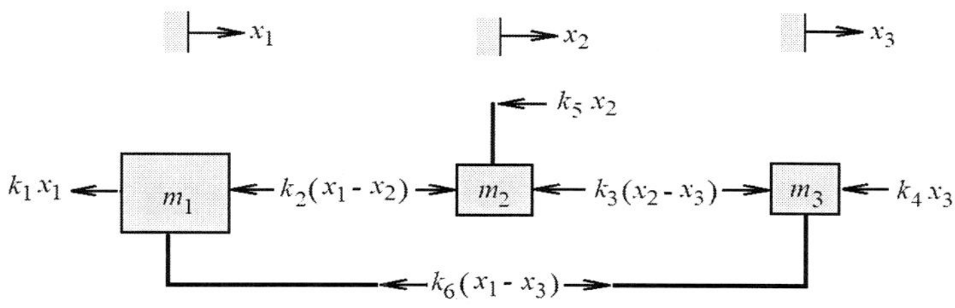

Figure 5.5a: Free body diagram of system in Figure 5.5.

Since the system has no damping, for oscillatory motion one can assume

$$x_1 = X_1 \cos \omega t, \quad x_2 = X_2 \cos \omega t, \quad x_3 = X_3 \cos \omega t$$

so that

$$\ddot{x}_1 = -\,\omega^2 X_1 \cos \omega t, \quad \ddot{x}_2 = -\,\omega^2 X_2 \cos \omega t, \quad \ddot{x}_3 = -\,\omega^2 X_3 \cos \omega t.$$

Substituting these values into Equations. (i) through (iii), it results in

$$(3k - m\omega^2)X_1 - kX_2 - kX_3 = 0, \qquad (iv)$$

$$-\,kX_1 + (3k - m\omega^2)X_2 - kX_3 = 0, \qquad (v)$$

$$-\,kX_1 - kX_2 + (3k - m\omega^2)X_3 = 0. \qquad (vi)$$

Writing in matrix form, one has

$$\begin{bmatrix} (3k - m\omega^2) & -\,k & -\,k \\ -\,k & (3k - m\omega^2) & -\,k \\ -\,k & -\,k & (3k - m\omega^2) \end{bmatrix} \begin{pmatrix} X_1 \\ X_2 \\ X_3 \end{pmatrix} = \begin{pmatrix} 0 \\ 0 \\ 0 \end{pmatrix}. \qquad (vii)$$

The frequency equation or characteristic equation is obtained by setting the determinant of the coefficient matrix to zero. That is,

$$\begin{vmatrix} (3k - m\omega^2) & -\,k & -\,k \\ -\,k & (3k - m\omega^2) & -\,k \\ -\,k & -\,k & (3k - m\omega^2) \end{vmatrix} = 0$$

which, upon operation, becomes

$$(3k - m\omega^2)^3 - 2k^3 - 3k^2(3k - m\omega^2) = 0. \qquad (viii)$$

But

$$(3k - m\omega^2)^3 = (3k - m\omega^2)(3k - m\omega^2)^2$$

$$= (3k - m\omega^2)(m^2\omega^4 - 6mk\omega^2 + 9k^2)$$

$$= 3k(m^2\omega^4 - 6mk\omega^2 + 9k^2) - m\omega^2(m^2\omega^4 - 6mk\omega^2 + 9k^2)$$

$$= 3k(m^2\omega^4 - 6mk\omega^2 + 9k^2) - m\omega^2(m^2\omega^4 - 6mk\omega^2 + 9k^2)$$

$$= 3km^2\omega^4 - 18mk^2\omega^2 + 27k^3 - m^3\omega^6 + 6m^2k\omega^4 - 9m\omega^2k^2$$

$$= -m^3\omega^6 + 9km^2\omega^4 - 27mk^2\omega^2 + 27k^3.$$

Substituting the above result into Equation $(viii)$, it results

$$-m^3\omega^6 + 9km^2\omega^4 - 27mk^2\omega^2 + 27k^3 - 2k^3 - 3k^2(3k - m\omega^2) = 0$$

$$-m^3\omega^6 + 9km^2\omega^4 - 27mk^2\omega^2 + 27k^3 - 2k^3 - 9k^3 + 3k^2m\omega^2 = 0$$

$$-m^3\omega^6 + 9km^2\omega^4 - 24mk^2\omega^2 + 16k^3 = 0$$

$$\omega^6 - 9\left(\frac{k}{m}\right)\omega^4 + 24\left(\frac{k}{m}\right)^2\omega^2 - 16\left(\frac{k}{m}\right)^3 = 0,$$

which gives

$$\left(\omega^2 - \frac{k}{m}\right)\left(\omega^2 - 4\frac{k}{m}\right)^2 = 0\,.$$

Therefore,

$$\omega_1^2 = \frac{k}{m}, \;\; \omega_2^2 = \omega_3^2 = 4\left(\frac{k}{m}\right).$$

Thus, the natural frequencies for the system are

$$\omega_1 = \sqrt{\frac{k}{m}}, \quad \omega_2 = \omega_3 = 2.0\sqrt{\frac{k}{m}}.$$

This 3dof system is *degenerate* because two of its natural frequencies are equal.

Example 5.4

The motions of two masses are confined to the plane as shown in Figure 5.6. For small oscillation, the motion in the horizontal direction is independent of that in the vertical direction. Derive the equations of motion and find the natural frequencies of the system. Given that $m_1 = m_2 = m$, $k_{x1} = k_{x3} = k_{y3} = 4k,\ k_{x2} = k_{y2} = k,\ k_{y1} = 3k,\ k_{y4} = 2k.$

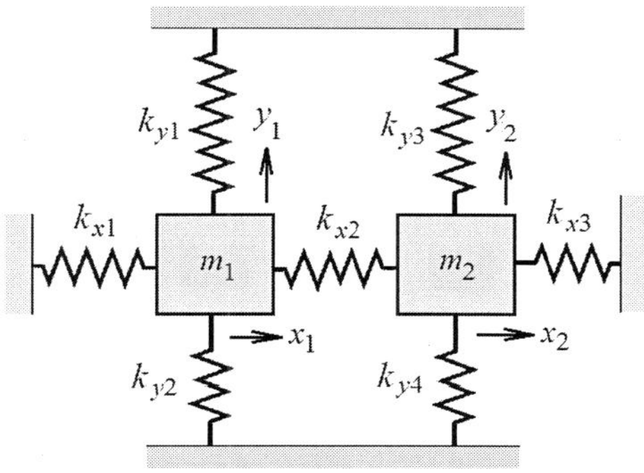

Figure 5.6: Plane motions of two spring-mass system.

Solution:

The equations of motion can be obtained by applying Newton's second law of motion for every direction.

Consider first motion along the horizontal direction. With reference to the FBD in Figure 5.6a, one obtains

$$m_1\ddot{x}_1 = -k_{x1}x_1 - k_{x2}(x_1 - x_2),$$

$$m_2\ddot{x}_2 = -k_{x3}x_2 + k_{x2}(x_1 - x_2)\,.$$

For motion along the vertical direction, one has

$$m_1\ddot{y}_1 = -\left(k_{y1} + k_{y2}\right)y_1,$$

$$m_2\ddot{y}_2 = -\left(k_{y3} + k_{y4}\right)y_2\,.$$

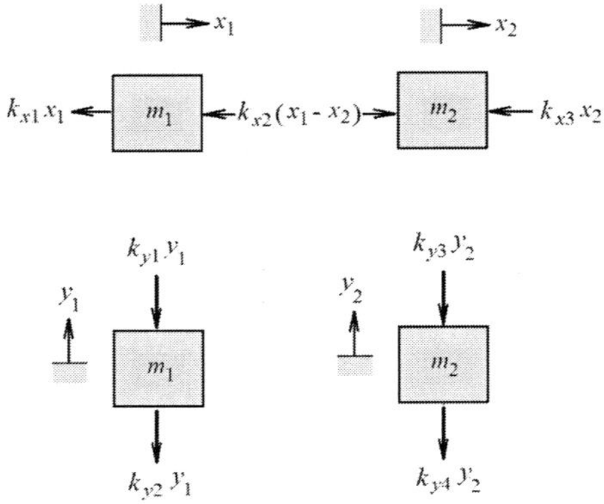

Figure 5.6a: FBD of system in Figure 5.6.

Rearranging terms,

$$m_1\ddot{x}_1 + (k_{x1} + k_{x2})x_1 - k_{x2}x_2 = 0,$$

$$m_2\ddot{x}_2 + (k_{x3} + k_{x2})x_2 - k_{x2}x_1 = 0,$$

$$m_1\ddot{y}_1 + (k_{y1} + k_{y2})y_1 = 0,$$

$$m_2\ddot{y}_2 + (k_{y3} + k_{y4})y_2 = 0.$$

After substituting for the given data, it leads to

$$m\ddot{x}_1 + 5kx_1 - kx_2 = 0, \qquad (i)$$

$$m\ddot{x}_2 + 5kx_2 - kx_1 = 0, \qquad (ii)$$

$$m\ddot{y}_1 + 4ky_1 = 0, \qquad (iii)$$

$$m\ddot{y}_2 + 6ky_2 = 0. \qquad (iv)$$

Writing these 4 equations in matrix form,

$$\begin{bmatrix} m & 0 & 0 & 0 \\ 0 & m & 0 & 0 \\ 0 & 0 & m & 0 \\ 0 & 0 & 0 & m \end{bmatrix} \begin{pmatrix} \ddot{x}_1 \\ \ddot{x}_2 \\ \ddot{y}_1 \\ \ddot{y}_2 \end{pmatrix} + \begin{bmatrix} 5k & -k & 0 & 0 \\ -k & 5k & 0 & 0 \\ 0 & 0 & 4k & 0 \\ 0 & 0 & 0 & 6k \end{bmatrix} \begin{pmatrix} x_1 \\ x_2 \\ y_1 \\ y_2 \end{pmatrix} = \begin{pmatrix} 0 \\ 0 \\ 0 \\ 0 \end{pmatrix}. \qquad (v)$$

This is the required matrix equation of motion for the given system.

Since the system has no damping and for oscillatory motion one can assume

$$x_1 = X_1 \cos \omega t,\quad x_2 = X_2 \cos \omega t,\quad y_1 = Y_1 \cos \omega t,\quad y_2 = Y_2 \cos \omega t$$

so that

$$\ddot{x}_1 = -\,\omega^2 X_1 \cos \omega t,\quad \ddot{x}_2 = -\,\omega^2 X_2 \cos \omega t,\quad \ddot{y}_1 = -\,\omega^2 Y_1 \cos \omega t,$$

$$\ddot{y}_1 = -\,\omega^2 Y_1 \cos \omega t,\qquad \ddot{y}_2 = -\,\omega^2 Y_2 \cos \omega t.$$

Substituting these values into Equation (v) and simplifying results in

$$\begin{bmatrix} (5k - m\omega^2) & -\,k & 0 & 0 \\ -\,k & (5k - m\omega^2) & 0 & 0 \\ 0 & 0 & (4k - m\omega^2) & 0 \\ 0 & 0 & 0 & (6k - m\omega^2) \end{bmatrix} \begin{pmatrix} X_1 \\ X_2 \\ Y_1 \\ Y_2 \end{pmatrix} = \begin{pmatrix} 0 \\ 0 \\ 0 \\ 0 \end{pmatrix}. \qquad (vi)$$

The frequency equation or characteristic equation is obtained by setting the determinant of the coefficient matrix to zero. That is,

$$\begin{vmatrix} (5k - m\omega^2) & -\,k & 0 & 0 \\ -\,k & (5k - m\omega^2) & 0 & 0 \\ 0 & 0 & (4k - m\omega^2) & 0 \\ 0 & 0 & 0 & (6k - m\omega^2) \end{vmatrix} = 0. \quad (vii)$$

With reference to Equation (v) and the fact that the horizontal motion is independent from that of the vertical motion, the system is in fact consisting of one 2dof system for the horizontal motion and two single dof systems for the two vertical motions. Thus, Equation (vii) can be split into three frequency equations as

$$\begin{vmatrix} (5k - m\omega^2) & -\,k \\ -\,k & (5k - m\omega^2) \end{vmatrix} = 0, \qquad 4k - m\omega^2 = 0,$$

$$6k - m\omega^2 = 0.$$

The first frequency equation gives

$$(5k - m\omega^2)^2 - k^2 = 0 \qquad \rightarrow \quad m^2\omega^4 - (10km)\,\omega^2 + 24k^2 = 0$$

which results in

$$\omega_1 = 2\sqrt{\frac{k}{m}}, \quad \omega_2 = \sqrt{\frac{6k}{m}}.$$

The second and the third frequency equations give

$$\omega_3 = 2\sqrt{\frac{k}{m}}, \quad \omega_4 = \sqrt{\frac{6k}{m}}.$$

In other words, the four natural frequencies of the given system are

$$\omega_1 = 2\sqrt{\frac{k}{m}}, \quad \omega_2 = \sqrt{\frac{6k}{m}}, \quad \omega_3 = 2\sqrt{\frac{k}{m}}, \quad \omega_4 = \sqrt{\frac{6k}{m}}.$$

5.3 Normal Modes, Orthogonality, and Expansion Theorem

In Section 4.5 the natural modes of a tdof system and its normalized form or normal modes as well as orthogonality property were briefly considered. In this section a more formal treatment is presented for a general linear discrete system having *n*dof. Specifically, Sub-section 5.3.1 includes normal modes and orthogonality while Sub-section 5.3.2 deals with the expansion theorem.

5.3.1 Normal modes and orthogonality

Consider a general linear discrete system whose matrix equation of motion is given by Equation (5.37). Since the system has no damping and for oscillatory motion one assume $x_i = X_i \cos \omega t$ such that $\ddot{x}_i = -\,\omega^2 X_i \cos \omega t$ in which $i = 1, 2, \dots, n$.

Substituting these values into Equation (5.37) and rearranging, one obtains

$$\begin{bmatrix} (k_{11}-m_{11}\omega^2) & k_{12} & \cdots & k_{1n} \\ k_{21} & (k_{22}-m_{22}\omega^2) & \cdots & k_{1n} \\ \cdots & \cdots & \ddots & \cdots \\ k_{n1} & k_{n2} & \cdots & (k_{nn}-m_{nn}\omega^2) \end{bmatrix} \begin{pmatrix} X_1 \\ X_2 \\ \cdots \\ X_n \end{pmatrix} = \begin{pmatrix} 0 \\ 0 \\ 0 \\ 0 \end{pmatrix} \quad \text{or}$$

$$[Z(\omega)](\vec{X}) = (\vec{0}) \tag{5.38}$$

where $[Z(\omega)]$ is the impedance matrix for the undamped system.

The solution for the natural frequencies and natural modes of the system is known as the *eigenvalue problem*. To obtain the natural frequencies of the system, one employs the frequency equation or characteristic equation which is found by setting the determinant of the coefficient matrix to zero. That is,

$$\begin{vmatrix} (k_{11}-m_{11}\omega^2) & k_{12} & \cdots & k_{1n} \\ k_{21} & (k_{22}-m_{22}\omega^2) & \cdots & k_{1n} \\ \cdots & \cdots & \ddots & \cdots \\ k_{n1} & k_{n2} & \cdots & (k_{nn}-m_{nn}\omega^2) \end{vmatrix} = 0.$$

From this equation the natural frequencies ω_i of the system are determined. Substituting ω_i one at time into the matrix equation above, one has

$$[Z(\omega_i)](\vec{X})_{(i)} = (\vec{0})$$

where $(\vec{X})_{(i)}$ is the i-natural mode which is a column vector and is called the i-*modal vector*. Usually, for convenience, $X_1 = 1$. The process of assigning a given value to a component of the modal vector is known as *normalization*. The *modal* or *transformation matrix* of the system is defined as

$$[R] = \left[\begin{pmatrix} X_1 \\ X_2 \\ \cdots \\ X_n \end{pmatrix}_{(1)} \begin{pmatrix} X_1 \\ X_2 \\ \cdots \\ X_n \end{pmatrix}_{(2)} \cdots \begin{pmatrix} X_1 \\ X_2 \\ \cdots \\ X_n \end{pmatrix}_{(n)} \right].$$

The modal vectors $\begin{pmatrix} X_1 \\ X_2 \\ \cdots \\ X_n \end{pmatrix}_{(i)}$ has a very useful property known as *orthogonality* with respect to the mass matrix $[M]$, and *orthogonality* with respect to the stiffness matrix $[K]$. It enables one to uncouple the mass and stiffness matrices such that every uncoupled equation of motion can be solved as a single dof system.

One can normalize the natural modes as that in Section 4.5 such that the normalized modal vectors are

$$\begin{pmatrix} \tilde{X}_1 \\ \tilde{X}_2 \\ \cdots \\ \tilde{X}_n \end{pmatrix}_{(i)}, \quad \begin{pmatrix} \tilde{X}_1 \\ \tilde{X}_2 \\ \cdots \\ \tilde{X}_n \end{pmatrix}_{(i)}^T [M] \begin{pmatrix} \tilde{X}_1 \\ \tilde{X}_2 \\ \cdots \\ \tilde{X}_n \end{pmatrix}_{(i)} = 1, \quad \begin{pmatrix} \tilde{X}_1 \\ \tilde{X}_2 \\ \cdots \\ \tilde{X}_n \end{pmatrix}_{(i)}^T [K] \begin{pmatrix} \tilde{X}_1 \\ \tilde{X}_2 \\ \cdots \\ \tilde{X}_n \end{pmatrix}_{(i)} = \omega_i^2, \text{ and}$$

$$\begin{pmatrix} \tilde{X}_1 \\ \tilde{X}_2 \\ \cdots \\ \tilde{X}_n \end{pmatrix}_{(i)}^T [M] \begin{pmatrix} \tilde{X}_1 \\ \tilde{X}_2 \\ \cdots \\ \tilde{X}_n \end{pmatrix}_{(s)} = 0, \quad \begin{pmatrix} \tilde{X}_1 \\ \tilde{X}_2 \\ \cdots \\ \tilde{X}_n \end{pmatrix}_{(i)}^T [K] \begin{pmatrix} \tilde{X}_1 \\ \tilde{X}_2 \\ \cdots \\ \tilde{X}_n \end{pmatrix}_{(s)} = 0, \quad s = 1, 2, \ldots, n$$

in which $i \neq s$. That is, the modal vectors $\begin{pmatrix} \tilde{X}_1 \\ \tilde{X}_2 \\ \cdots \\ \tilde{X}_n \end{pmatrix}_{(i)}$ and $\begin{pmatrix} \tilde{X}_1 \\ \tilde{X}_2 \\ \cdots \\ \tilde{X}_n \end{pmatrix}_{(s)}$ are *orthogonal.*

The solutions for natural frequencies and normal modes of several *n*dof systems are presented in Sections 5.4 and 5.5.

5.3.2 Expansion theorem

The modal vectors or eigenvectors, due their orthogonal property, are linearly independent. Therefore, they provide a basis in the *n*-dimensional (*n*D) space. This implies that any vector in the *n*D space can be written as a linear combination of the *n* linearly independent vectors. In other words, if $\vec{x}$ is any vector in the *n*D space it can be written as

$$(\vec{x}) = \sum_{i=1}^{n} \alpha_i \begin{pmatrix} X_1 \\ X_2 \\ \cdots \\ X_n \end{pmatrix}_{(i)} \tag{5.39}$$

where α_i are constants and can be evaluated by premultiplying this equation throughout with $\begin{pmatrix} X_1 \\ X_2 \\ \cdots \\ X_n \end{pmatrix}_{(i)}^T [M]$ and applying the orthogonality property such that

$$\alpha_i \left[\begin{pmatrix} X_1 \\ X_2 \\ \cdots \\ X_n \end{pmatrix}_{(i)}^T [M] \begin{pmatrix} X_1 \\ X_2 \\ \cdots \\ X_n \end{pmatrix}_{(i)} \right] = \begin{pmatrix} X_1 \\ X_2 \\ \cdots \\ X_n \end{pmatrix}_{(i)}^T [M] (\vec{x}) .$$

If the normalized modal vectors are employed instead,

$$\alpha_i = \frac{\begin{pmatrix} \tilde{X}_1 \\ \tilde{X}_2 \\ \cdots \\ \tilde{X}_n \end{pmatrix}_{(i)}^T [M] \vec{x}}{\begin{pmatrix} \tilde{X}_1 \\ \tilde{X}_2 \\ \cdots \\ \tilde{X}_n \end{pmatrix}_{(i)}^T [M] \begin{pmatrix} \tilde{X}_1 \\ \tilde{X}_2 \\ \cdots \\ \tilde{X}_n \end{pmatrix}} = \begin{pmatrix} \tilde{X}_1 \\ \tilde{X}_2 \\ \cdots \\ \tilde{X}_n \end{pmatrix}_{(i)}^T [M](\vec{x}) \qquad i = 1, 2, \dots, n. \quad (5.40)$$

Equation (5.39) is known as the *expansion theorem* (Amirouche, 1992, 187). It forms the basis of the method of obtaining the system response in term of normal modes of vibration. The latter method is known as *modal analysis*.

5.4 Normal Mode Vibration and Response of Undamped Systems

In this section a specific system is considered so as to illustrate application of the procedure presented in the last section.

A system of three identical blocks of mass rests on a frictionless surface as shown in Figure 5.7. Every block having mass of $m = 1$ kg is connected by springs of coefficient of stiffness $k = 1$ N/m. If block m_3 is given an initial displacement of 1 m, determine the resulting motion of the system.

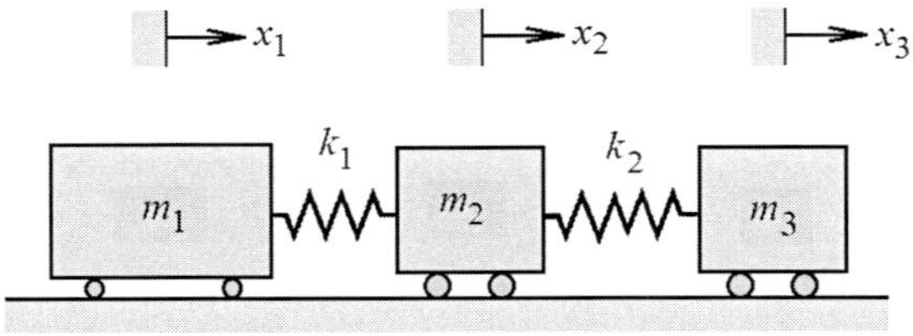

Figure 5.7: A system of mass-springs with 3dof.

Solution:
For generality, one starts writing the three identical masses as $m_1 = m_2 = m_3 = m$ and the two identical stiffness as $k_1 = k_2 = k$.

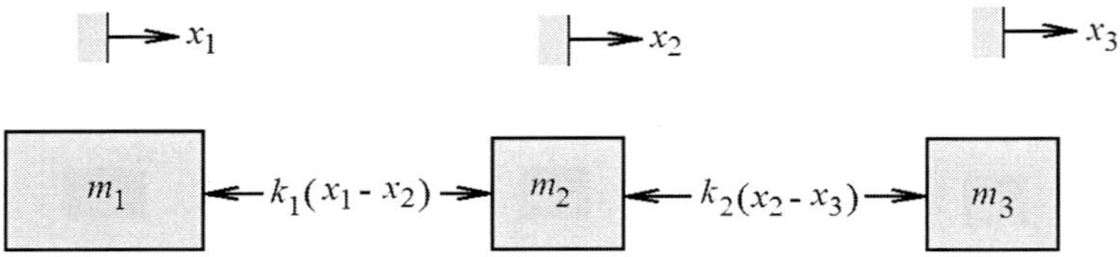

Figure 5.7a: FBD of system in Figure 5.7.

The equations of motion can be obtained by applying Newton's second law of motion, $\sum F = ma$ to every mass. By making use of the FBD as shown in Figure 5.7a, one has

$$m_1\ddot{x}_1 = -k_1(x_1 - x_2),$$

$$m_2\ddot{x}_2 = k_1(x_1 - x_2) - k_2(x_2 - x_3),$$

$$m_3\ddot{x}_3 = k_2(x_2 - x_3).$$

Moving all the terms to the lhs of the equality sign, one has

$$m_1\ddot{x}_1 + k_1(x_1 - x_2) = 0,$$

$$m_2\ddot{x}_2 - k_1(x_1 - x_2) + k_2(x_2 - x_3) = 0,$$

$$m_3\ddot{x}_3 + k_2(x_3 - x_2) = 0.$$

After substituting for the given data, one obtains

$$\ddot{x}_1 + x_1 - x_2 = 0, \qquad (i)$$

$$\ddot{x}_2 + 2x_2 - x_1 - x_3 = 0, \qquad (ii)$$

$$\ddot{x}_3 + x_3 - x_2 = 0. \qquad (iii)$$

(a) ***Natural frequencies of system***
Since the system has no damping and therefore there is no delay such that for oscillatory motion one can assume

$$x_1 = X_1 \cos \omega t, \quad x_2 = X_2 \cos \omega t, \quad x_3 = X_3 \cos \omega t$$

so that

$$\ddot{x}_1 = -\omega^2 X_1 \cos \omega t, \quad \ddot{x}_2 = -\omega^2 X_2 \cos \omega t, \quad \ddot{x}_3 = -\omega^2 X_3 \cos \omega t.$$

$$\ddot{x}_1 + x_1 - x_2 = 0, \qquad (i)$$

$$\ddot{x}_2 + 2x_2 - x_1 - x_3 = 0, \qquad (ii)$$

$$\ddot{x}_3 + x_3 - x_2 = 0. \qquad (iii)$$

Substituting these values into Equations (i) through (iii), it results in

$$(1 - \omega^2)X_1 - X_2 = 0, \qquad (iv)$$

$$-X_1 + (2 - \omega^2)X_2 - X_3 = 0, \qquad (v)$$

$$-X_2 + (1 - \omega^2)X_3 = 0. \qquad (vi)$$

Writing in matrix form, one has

$$\begin{bmatrix} (1-\omega^2) & -1 & 0 \\ -1 & (2-\omega^2) & -1 \\ 0 & -1 & (1-\omega^2) \end{bmatrix} \begin{pmatrix} X_1 \\ X_2 \\ X_3 \end{pmatrix} = \begin{pmatrix} 0 \\ 0 \\ 0 \end{pmatrix}. \qquad (vii)$$

The frequency equation or characteristic equation is obtained by setting the determinant of the coefficient matrix to zero. That is,

$$\begin{vmatrix} (1-\omega^2) & -1 & 0 \\ -1 & (2-\omega^2) & -1 \\ 0 & -1 & (1-\omega^2) \end{vmatrix} = 0$$

which, upon operation, becomes

$$(1-\omega^2)^2(2-\omega^2) - 2(1-\omega^2) = 0$$

$$(1-\omega^2)[(2-\omega^2)(1-\omega^2) - 2] = 0$$

$$(1-\omega^2)(2 - 2\omega^2 - \omega^2 + \omega^4 - 2) = 0$$

$$(1-\omega^2)\omega^2(\omega^2 - 3) = 0$$

which gives

$$\omega_1 = 0\,, \quad \omega_2 = 1 \ \text{rad/s}\,, \quad \omega_3 = \sqrt{3} \ \text{rad/s}.$$

(b) ***Mode shapes of system***

To find the mode shapes or eigenvectors of the system one substitutes the natural frequencies

$$\omega_1 = 0\,, \quad \omega_2 = 1 \ \text{rad/s}\,, \quad \omega_3 = \sqrt{3} \ \text{rad/s}$$

into Equation (vii) or Equations (iv) through (vi) and solving for the mode shapes $\vec{X}_{(1)}$, $\vec{X}_{(2)}$, $\vec{X}_{(3)}$, respectively. Thus, the mode shape associated with the first natural frequency is given as

$$\vec{X}_{(1)} = \begin{pmatrix} X_1 \\ X_2 \\ X_3 \end{pmatrix}_{(1)}$$

where the subscript (1) denotes the vector associated with ω_1, and so on. Thus, for $\omega_1 = 0$, Equation (iv) provides

$$X_{1(1)} - X_{2(1)} = 0,$$

Equation (v) gives

$$-X_{1(1)} + 2X_{2(1)} - X_{3(1)} = 0,$$

$$-X_{2(1)} + X_{3(1)} = 0.$$

As there are 3 unknowns but there are only two independent equations therefore, one can arbitrarily select a value for any of the 3 unknowns in order to solve for these 3 equations.

For convenience and simplicity, one can assume $X_{1(1)} = 1$ so that the above 3 equations lead to

$$X_{1(1)} = 1, \qquad X_{2(1)} = 1, \qquad X_{3(1)} = 1$$

or in matrix form

$$\vec{X}_{(1)} = \begin{pmatrix} X_1 \\ X_2 \\ X_3 \end{pmatrix}_{(1)} = \begin{pmatrix} X_{1(1)} \\ X_{2(1)} \\ X_{3(1)} \end{pmatrix} = \begin{pmatrix} 1 \\ 1 \\ 1 \end{pmatrix}. \qquad (viii)$$

Similarly, the mode shape associated with the second natural frequency $\omega_2 = 1 \text{ rad/s}$, and assuming $X_{1(2)} = 1$, becomes

$$\vec{X}_{(2)} = \begin{pmatrix} X_1 \\ X_2 \\ X_3 \end{pmatrix}_{(2)} = \begin{pmatrix} X_{1(2)} \\ X_{2(2)} \\ X_{3(2)} \end{pmatrix} = \begin{pmatrix} 1 \\ 0 \\ -1 \end{pmatrix}. \qquad (ix)$$

The mode shape associated with the third natural frequency $\omega_3 = \sqrt{3} \text{ rad/s}$, and assuming $X_{1(3)} = 1$, becomes

$$\vec{X}_{(3)} = \begin{pmatrix} X_1 \\ X_2 \\ X_3 \end{pmatrix}_{(3)} = \begin{pmatrix} X_{1(3)} \\ X_{2(3)} \\ X_{3(3)} \end{pmatrix} = \begin{pmatrix} 1 \\ -2 \\ 1 \end{pmatrix}. \qquad (x)$$

(c) ***Responses of system***

In general, the responses of the system are provided by the Expansion Theorem, Equation (5.39), as

$$\vec{x} = \vec{x}(t) = \begin{pmatrix} x_1(t) \\ x_2(t) \\ x_3(t) \end{pmatrix} = \sum_{i=1}^{3} \vec{X}_{(i)} X_i \cos(\omega_i t + \varphi_i)$$

$$= \left[\begin{pmatrix} X_1 \\ X_2 \\ X_3 \end{pmatrix}_{(1)} \begin{pmatrix} X_1 \\ X_2 \\ X_3 \end{pmatrix}_{(2)} \begin{pmatrix} X_1 \\ X_2 \\ X_3 \end{pmatrix}_{(3)} \right] \begin{pmatrix} X_1 \cos(\omega_1 t + \varphi_1) \\ X_2 \cos(\omega_2 t + \varphi_2) \\ X_3 \cos(\omega_3 t + \varphi_3) \end{pmatrix} \quad (xi)$$

in which the coefficient matrix is known as the modal matrix whose columns are the mode shapes or eigenvectors of the system. The amplitudes X_i and phase angles or delays φ_i are constants and have to be determined by applying the initial conditions of the system.

In the present system there is no damping and therefore there is no phase angle or delay so that upon substituting the mode shapes the responses reduce to

$$\begin{pmatrix} x_1(t) \\ x_2(t) \\ x_3(t) \end{pmatrix} = \begin{bmatrix} 1 & 1 & 1 \\ 1 & 0 & -2 \\ 1 & -1 & 1 \end{bmatrix} \begin{pmatrix} X_1 \cos \omega_1 t \\ X_2 \cos \omega_2 t \\ X_3 \cos \omega_3 t \end{pmatrix}. \qquad (xii)$$

Applying the given initial conditions, it becomes

$$\begin{pmatrix} x_1(0) \\ x_2(0) \\ x_3(0) \end{pmatrix} = \begin{pmatrix} 0 \\ 0 \\ 1 \end{pmatrix} = \begin{bmatrix} 1 & 1 & 1 \\ 1 & 0 & -2 \\ 1 & -1 & 1 \end{bmatrix} \begin{pmatrix} X_1 \\ X_2 \\ X_3 \end{pmatrix}.$$

This gives

$$X_1 = \frac{1}{3}\text{ m}, \quad X_2 = \frac{1}{2}\text{ m}, \ , \quad X_3 = \frac{1}{6}\text{ m}.$$

Substituting these values and the natural frequencies into Equation (xii), one has

$$\begin{pmatrix} x_1(t) \\ x_2(t) \\ x_3(t) \end{pmatrix} = \begin{bmatrix} 1 & 1 & 1 \\ 1 & 0 & -2 \\ 1 & -1 & 1 \end{bmatrix} \begin{pmatrix} \frac{1}{3}\cos 0t \\ \frac{1}{2}\cos t \\ \frac{1}{6}\cos\sqrt{3}\,t \end{pmatrix} \quad \text{m}$$

or expanding the matrix equation to give

$$x_1(t) = \frac{1}{3} + \frac{1}{2}\cos t + \frac{1}{6}\cos\sqrt{3}\,t \quad \text{m},$$

$$x_2(t) = \frac{1}{3} - \frac{1}{3}\cos\sqrt{3}\,t \quad \text{m},$$

$$x_3(t) = \frac{1}{3} - \frac{1}{2}\cos t + \frac{1}{6}\cos\sqrt{3}\,t \quad \text{m}.$$

The constant term in every response is due to the fact that the system is not constrained and therefore it is due to the rigid-body mode $\vec{X}_{(1)}$.

5.5 Responses of Forced Mdof Systems with Proportional Damping

As mentioned in Section 4.6, for systems with proportional damping, if more than two independent modal damping parameters ζ_r are to be specified the uncoupling condition in Equation (4.28) has to be modified in accordance with the following relation

$$[C] = [M] \sum_{p=0} \lambda_p \, [M^{-1}K]^p \tag{5.41}$$

where λ_p are constants to be obtained and $p = 0, 1, 2, ...$ The upper limit of the summation is not defined as it varies from one system to another.

The governing matrix equation of motion in terms of the generalized displacement, velocity, acceleration and forcing vectors is given by

$$[M](\ddot{\vec{x}}) + [C\}(\dot{\vec{x}}) + [K](\vec{x}) = (\vec{F}) \tag{5.42}$$

where $(\vec{F})$ is the vector of generalized applied forces. With the definition of damping matrix $[C]$ in Equation (5.41), the following algebraic equations (Rao, 1995, 431) can be applied to determine constants λ_p,

$$2\zeta_r \omega_r = \sum_{p=0} \lambda_p \, \omega_r^{2p} \qquad r = 1, 2, ..., n. \tag{5.43}$$

Many equations in Equation (5.40) have to be included as there are specified modal damping ratios.

The foregoing way of forming the damping matrix $[C]$ seems to be straightforward. However, when more modes are considered in Equation (5.43), it yields a damping matrix that is not diagonal after the normal mode transformation is performed. The consequence of this is that computational difficulty is likely to occur.

Example 5.5

The 3dof linear system shown in Figure 5.8 has modal damping parameters, $\zeta_r = 0.01, \quad r = 1, 2, 3.$

(a) Derive the matrix equation of motion for the system.

(b) Find the undamped natural frequencies and mode shapes of the system.

(c) Form the damping matrix $[C] = [M] \sum_{p=0} \lambda_p [M^{-1}K]^p$ so that the modal damping parameters become $2\zeta_r \omega_r = \sum_{p=0} \lambda_p \omega_r^{2p}, \quad r = 1, 2, 3.$

(d) Determine the steady-state response of the system shown in Figure 5.8 in which

$$F_1(t) = F_2(t) = F_3(t) = F \cos \omega t, \quad \omega = \frac{7}{4}\sqrt{\frac{k}{m}}.$$

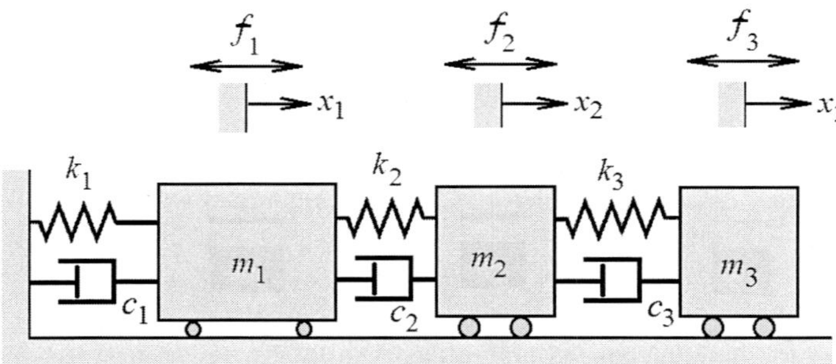

Figure 5.8: 3dof linear system with damping.

Solution*:*

(a) ***The equations of motion***

Applying the FBD in Figure 5.8a, for the mass at the top, the equations of motion can be obtained by applying Newton's second law of motion, $\sum F = ma$ to every mass. By making use of the FBD as shown in Figure 5.8a, one has for the top mass,

$$m\ddot{x}_1 = -kx_1 - k(x_1 - x_2) - c_1\dot{x}_1 - c_2(\dot{x}_1 - \dot{x}_2) + f_1(t).$$

Rearranging,

$$m\ddot{x}_1 + (c_1 + c_2)\dot{x}_1 - c_2\dot{x}_2 + 2kx_1 - kx_2 = f_1(t).$$

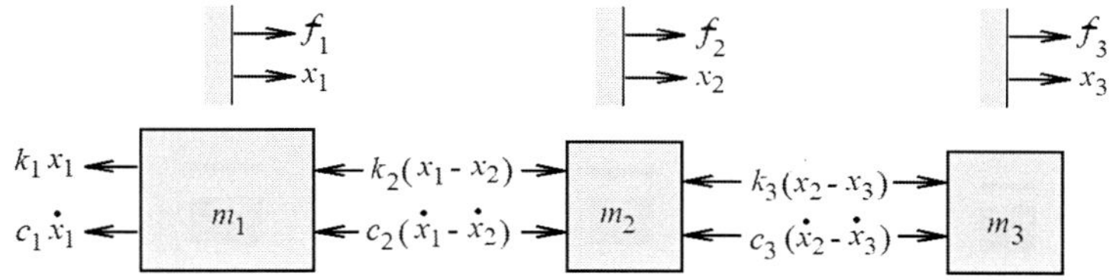

Figure 5.8a: FBD of linear system in Figure 5.8.

For the middle mass, one obtain

$$m\ddot{x}_2 = k(x_1 - x_2) - k(x_2 - x_3) + c_2(\dot{x}_1 - \dot{x}_2) - c_3(\dot{x}_2 - \dot{x}_3) + f_2(t).$$

Rearranging,

$$m\ddot{x}_2 - c_2\dot{x}_1 + (c_2 + c_3)\dot{x}_2 - c_3\dot{x}_3 - kx_1 + 2kx_2 - kx_3 = f_2(t).$$

For the lower mass, one has

$$m\ddot{x}_3 = k(x_2 - x_3) + c_3(\dot{x}_2 - \dot{x}_3) + f_3(t).$$

Rearranging,

$$m\ddot{x}_3 - c_3\dot{x}_2 + c_3\dot{x}_3 - kx_2 + kx_3 = f_3(t).$$

The matrix equation of motion becomes

$$[M](\ddot{\vec{x}}) + [C\}(\dot{\vec{x}}) + [K](\vec{x}) = (\vec{F})$$

in which the mass, damping, and stiffness matrices are

$$[M] = m\begin{bmatrix} 1 & 0 & 0 \\ 0 & 1 & 0 \\ 0 & 0 & 1 \end{bmatrix}, \qquad [C] = \begin{bmatrix} (c_1 + c_2) & -c_2 & 0 \\ -c_2 & (c_2 + c_3) & -c_3 \\ 0 & -c_3 & c_3 \end{bmatrix},$$

$$[K] = k\begin{bmatrix} 2 & -1 & 0 \\ -1 & 2 & -1 \\ 0 & -1 & 1 \end{bmatrix}.$$

(b) ***Natural frequencies and mode shapes of system***
For the undamped natural frequencies, one can disregard the damping matrix such that one can assume

$$x_1 = X_1 \cos \omega t, \quad x_2 = X_2 \cos \omega t, \quad x_3 = X_3 \cos \omega t$$

so that

$$\ddot{x}_1 = -\,\omega^2 X_1 \cos \omega t, \quad \ddot{x}_2 = -\,\omega^2 X_2 \cos \omega t, \quad \ddot{x}_3 = -\,\omega^2 X_3 \cos \omega t.$$

The undamped matrix equation of motion is

$$[M](\ddot{\vec{x}}\,) + [K](\vec{x}) = (\vec{0})$$

or upon expanding

$$m\ddot{x}_1 + 2kx_1 - kx_2 = 0, \qquad (i)$$

$$m\ddot{x}_2 - kx_1 + 2kx_2 - kx_3 = 0, \qquad (ii)$$

$$m\ddot{x}_3 - kx_2 + kx_3 = 0. \qquad (iii)$$

For the undamped natural frequencies, one can assume

$$x_1 = X_1 \cos \omega t, \quad x_2 = X_2 \cos \omega t, \quad x_3 = X_3 \cos \omega t$$

so that

$$\ddot{x}_1 = -\,\omega^2 X_1 \cos \omega t, \quad \ddot{x}_2 = -\,\omega^2 X_2 \cos \omega t, \quad \ddot{x}_3 = -\,\omega^2 X_3 \cos \omega t.$$

Substituting these values into Eqs. (i) through (iii), they become

$$(2k - m\omega^2)X_1 - kX_2 = 0, \qquad (iv)$$

$$-kX_1 + (2k - m\omega^2)X_2 - kX_3 = 0, \qquad (v)$$

$$-\,kX_2 + (k - m\omega^2)X_3 = 0. \qquad (vi)$$

Writing in matrix form, one has

$$\begin{bmatrix} (2k - m\omega^2) & -\,k & 0 \\ -\,k & (2k - m\omega^2) & -\,k \\ 0 & -\,k & (k - m\omega^2) \end{bmatrix} \begin{pmatrix} X_1 \\ X_2 \\ X_3 \end{pmatrix} = \begin{pmatrix} 0 \\ 0 \\ 0 \end{pmatrix}. \qquad (vii)$$

The frequency equation or characteristic equation is obtained by setting the determinant of the coefficient matrix to zero. That is,

$$\begin{vmatrix} (2k - m\omega^2) & -k & 0 \\ -k & (2k - m\omega^2) & -k \\ 0 & -k & (k - m\omega^2) \end{vmatrix} = 0$$

which, upon operating, becomes

$$(2k - m\omega^2)(2k - m\omega^2)(k - m\omega^2) - (2k - m\omega^2)k^2 - (k - m\omega^2)k^2 = 0.$$

Writing $\lambda = \frac{m\omega^2}{k}$, the above equation reduces to

$$k(2 - \lambda)k(2 - \lambda)k(1 - \lambda) - k(2 - \lambda)k^2 - k(1 - \lambda)k^2 = 0$$

$$(2 - \lambda)(2 - \lambda)(1 - \lambda) - (2 - \lambda) - (1 - \lambda) = 0$$

$$\lambda^3 - 5\lambda^2 + 6\lambda - 1 = 0.$$

This cubic equation gives

$$\lambda_1 = \frac{m\omega_1^2}{k} = 0.1981\,, \lambda_2 = \frac{m\omega_2^2}{k} = 1.5550, \lambda_3 = \frac{m\omega_3^2}{k} = 3.2470.$$

Therefore, the undamped natural frequencies of the system are

$$\omega_1 = 0.4451\sqrt{\frac{k}{m}}, \qquad \omega_2 = 1.2470\sqrt{\frac{k}{m}}, \qquad \omega_3 = 1.8019\sqrt{\frac{k}{m}}.$$

In order to obtain the mode shapes of the system the natural frequencies are substituted into Equation (vii), one at a time. To simplify the algebraic manipulation, one divides Equation (vii) by k so that the resulting equation becomes

$$\begin{bmatrix} \left(2-\frac{m\omega^2}{k}\right) & -1 & 0 \\ -1 & \left(2-\frac{m\omega^2}{k}\right) & -1 \\ 0 & -1 & \left(1-\frac{m\omega^2}{k}\right) \end{bmatrix} \begin{pmatrix} X_1 \\ X_2 \\ X_3 \end{pmatrix} = \begin{pmatrix} 0 \\ 0 \\ 0 \end{pmatrix}. \qquad (viii)$$

For ω_1, $\frac{m\omega_1^2}{k} = 0.1981$ such that Equation $(viii)$ results in

$$\begin{bmatrix} (2-0.1981) & -1 & 0 \\ -1 & (2-0.1981) & -1 \\ 0 & -1 & (1-0.1981) \end{bmatrix} \begin{pmatrix} X_1 \\ X_2 \\ X_3 \end{pmatrix}_{(1)} = \begin{pmatrix} 0 \\ 0 \\ 0 \end{pmatrix}.$$

$$\begin{bmatrix} 1.8019 & -1 & 0 \\ -1 & 1.8019 & -1 \\ 0 & -1 & 0.8019 \end{bmatrix} \begin{pmatrix} X_1 \\ X_2 \\ X_3 \end{pmatrix}_{(1)} = \begin{pmatrix} 0 \\ 0 \\ 0 \end{pmatrix}.$$

Expanding,

$$1.8019\, X_{1(1)} - X_{2(1)} = 0$$

$$-X_{1(1)} + 1.8019\, X_{2(1)} - X_{3(1)} = 0$$

$$-X_{2(1)} + 0.8019 X_{3(1)} = 0.$$

Subtracting the third equation from the first, one has

$$1.8019\, X_{1(1)} - 0.8019 X_{3(1)} = 0.$$

From the first equation, $X_{2(1)} = 1.8019\, X_{1(1)}$. Since there are two independent equations but there are three variables. Thus, one can choose $X_{1(1)} = 1$ so that $X_{2(1)} = 1.8019$. Substituting these values into the second equation, one finds

$$X_{3(1)} = -1 + 1.8019(1.8019) = 2.2468$$

which gives the first mode shape as

$$\vec{X}_{(1)} = \begin{pmatrix} X_1 \\ X_2 \\ X_3 \end{pmatrix}_{(1)} = \begin{pmatrix} X_{1(1)} \\ X_{2(1)} \\ X_{3(1)} \end{pmatrix} = \begin{pmatrix} 1 \\ 1.8019 \\ 2.2468 \end{pmatrix}. \qquad (ix)$$

Similarly, for ω_2, $\frac{m\omega_2^2}{k} = 1.5550$ such that Equation $(viii)$ results in

$$\begin{bmatrix} (2-1.5550) & -1 & 0 \\ -1 & (2-1.5550) & -1 \\ 0 & -1 & (1-1.5550) \end{bmatrix} \begin{pmatrix} X_1 \\ X_2 \\ X_3 \end{pmatrix}_{(2)} = \begin{pmatrix} 0 \\ 0 \\ 0 \end{pmatrix}.$$

$$\begin{bmatrix} 0.4450 & -1 & 0 \\ -1 & 0.4450 & -1 \\ 0 & -1 & -0.5550 \end{bmatrix} \begin{pmatrix} X_1 \\ X_2 \\ X_3 \end{pmatrix}_{(2)} = \begin{pmatrix} 0 \\ 0 \\ 0 \end{pmatrix}.$$

Expanding,

$$0.4450\, X_{1(2)} - X_{2(2)} = 0$$

$$-X_{1(2)} + 0.4450\, X_{2(2)} - X_{3(2)} = 0$$

$$-X_{2(2)} - 0.5550 X_{3(2)} = 0.$$

Subtracting the third equation from the first, one has

$$0.4450\, X_{1(2)} + 0.5550 X_{3(12)} = 0.$$

From the first equation, $X_{2(2)} = 0.4450\, X_{1(2)}$. Since there are two independent equations but there are three variables. Thus, one can chose $X_{1(2)} = 1$ so that $X_{2(2)} = 0.4450$. Substituting these values into the second equation, one finds

$$X_{3(2)} = -1 + 0.4450(0.4450) = -0.8020$$

which gives the second mode shape as

$$\vec{X}_{(2)} = \begin{pmatrix} X_1 \\ X_2 \\ X_3 \end{pmatrix}_{(2)} = \begin{pmatrix} X_{1(2)} \\ X_{2(2)} \\ X_{3(2)} \end{pmatrix} = \begin{pmatrix} 1 \\ 0.4450 \\ -0.8020 \end{pmatrix}. \qquad (x)$$

Now, for ω_3, $\frac{m\omega_3^2}{k} = 3.2470$ such that Equation $(viii)$ results in

$$\begin{bmatrix} (2-3.2470\ ) & -1 & 0 \\ -1 & (2-3.2470\ ) & -1 \\ 0 & -1 & (1-3.2470\ ) \end{bmatrix} \begin{pmatrix} X_1 \\ X_2 \\ X_3 \end{pmatrix}_{(3)} = \begin{pmatrix} 0 \\ 0 \\ 0 \end{pmatrix}.$$

$$\begin{bmatrix} -1.2470 & -1 & 0 \\ -1 & -1.2470 & -1 \\ 0 & -1 & -2.2470 \end{bmatrix} \begin{pmatrix} X_1 \\ X_2 \\ X_3 \end{pmatrix}_{(3)} = \begin{pmatrix} 0 \\ 0 \\ 0 \end{pmatrix}.$$

Expanding,

$$-1.2470\, X_{1(3)} - X_{2(3)} = 0$$

$$-X_{1(3)} - 1.2470\, X_{2(3)} - X_{3(3)} = 0$$

$$-X_{2(3)} - 2.2470 X_{3(3)} = 0.$$

Subtracting the third equation from the first, one has

$$-1.2470\, X_{1(3)} + 2.2470 X_{3(3)} = 0.$$

From the first equation, $X_{2(3)} = -1.2470\, X_{1(3)}$. Since there are two independent equations but there are three variables. Thus, one can chose $X_{1(3)} = 1$ so that $X_{2(3)} = -1.2470$. Substituting these values into the second equation, one finds

$$X_{3(3)} = -1 + 1.2470(1.2470) = 0.5550$$

which gives the second mode shape as

$$\vec{X}_{(3)} = \begin{pmatrix} X_1 \\ X_2 \\ X_3 \end{pmatrix}_{(3)} = \begin{pmatrix} X_{1(3)} \\ X_{2(3)} \\ X_{3(3)} \end{pmatrix} = \begin{pmatrix} 1 \\ -1.2470 \\ 0.5550 \end{pmatrix}. \qquad (xi)$$

Equations $(ix), (x),$ and (xi) give the three mode shapes of the system.

(c) ***Construction of proportional damping matrix*** $[\boldsymbol{C}]$

The damping matrix $[C] = [M]\sum_{p=0} \lambda_p\, [M^{-1}K]^p$ for which viscous damping of the proportional type with damping parameters defined by

$2\zeta_r\omega_r = \sum_{p=0} \lambda_p\, \omega_r^{2p}$, $r = 1, 2, 3$ with three independent ζ_r can be constructed. In other words, for three independent ζ_r to be specified and the damping matrix can be uncoupled there should be three λ_p with $p = 0, 1, 2$. With $p = 0, 1$ the third damping parameter is related to the first and second. Thus, it may be appropriate to mention that the example in (Clough, 1975, 195) is incorrect in the sense that the damping matrix defined there, namely $[C] = \alpha[M] + \beta[K]$ cannot be uncoupled since the three values of ζ_r were independently specified.

To obtain the λ_p with $p = 0, 1, 2$ for the system, one can write

$$2\zeta_1\omega_1 = \lambda_o + \lambda_1\omega_1^2 + \lambda_2\omega_1^4,$$

$$2\zeta_2\omega_2 = \lambda_o + \lambda_1\omega_2^2 + \lambda_2\omega_2^4,$$

$$2\zeta_3\omega_3 = \lambda_o + \lambda_1\omega_3^2 + \lambda_2\omega_3^4.$$

Solving these three equations for λ_o, λ_1, λ_2 and the construction of the proportional damping matrix in accordance with Equation (5.41), then the matrix equation of motion derived in part (a) above, with the derived damping matrix there to be replaced by the newly constructed proportional damping matrix, can be uncoupled.

(d) ***Steady-state response of the system***
Consider now the equations of motion

$$[M](\ddot{\vec{x}}) + [C\}(\dot{\vec{x}}) + [K](\vec{x}) = (\vec{F}) \qquad \text{or}$$

$$[M]\,[R](\ddot{\vec{q}}) + [C\}[R](\dot{\vec{q}}) + [K][R](\vec{q}) = (\vec{F}) \qquad (xii)$$

where $(\vec{x}) = [R](\vec{q})$, the transformation matrix $[R]$ is defined by

$$[R] = \left[\begin{pmatrix} X_{1(1)} \\ X_{2(1)} \\ X_{3(1)} \end{pmatrix} \begin{pmatrix} X_{1(2)} \\ X_{2(2)} \\ X_{3(2)} \end{pmatrix} \begin{pmatrix} X_{1(3)} \\ X_{2(3)} \\ X_{3(3)} \end{pmatrix}\right],$$

$$[R] = \left[\begin{pmatrix} 1 \\ 1.8019 \\ 2.2468 \end{pmatrix} \begin{pmatrix} 1 \\ 0.4450 \\ -0.8020 \end{pmatrix} \begin{pmatrix} 1 \\ -1.2470 \\ 0.5550 \end{pmatrix}\right].$$

Pre-multiplying Equation (xii) by $[R]^T$, one obtains

$$[R]^T[M]\,[R](\ddot{\vec{q}}\,) + [R]^T[C\}[R](\dot{\vec{q}}) + \;[R]^T[K][R](\vec{q}) = [R]^T(\vec{F}) \quad (xiii)$$

Before proceeding further to the solution of the system, one can normalize the natural modes or mode shapes. To this end, one let the normalized natural modes be

$$\begin{pmatrix} \tilde{X}_{1(1)} \\ \tilde{X}_{2(1)} \\ \tilde{X}_{3(1)} \end{pmatrix}, \quad \begin{pmatrix} \tilde{X}_{1(2)} \\ \tilde{X}_{2(2)} \\ \tilde{X}_{3(2)} \end{pmatrix}, \quad \begin{pmatrix} \tilde{X}_{1(3)} \\ \tilde{X}_{2(3)} \\ \tilde{X}_{3(3)} \end{pmatrix}$$

such that

$$\begin{pmatrix} \tilde{X}_{1(1)} \\ \tilde{X}_{2(1)} \\ \tilde{X}_{3(1)} \end{pmatrix} = \mu_1 \begin{pmatrix} X_{1(1)} \\ X_{2(1)} \\ X_{3(1)} \end{pmatrix} = \mu_1 \begin{pmatrix} 1 \\ 1.8019 \\ 2.2468 \end{pmatrix},$$

$$\begin{pmatrix} \tilde{X}_{1(2)} \\ \tilde{X}_{2(2)} \\ \tilde{X}_{3(2)} \end{pmatrix} = \mu_2 \begin{pmatrix} X_{1(2)} \\ X_{2(2)} \\ X_{3(2)} \end{pmatrix} = \mu_2 \begin{pmatrix} 1 \\ 0.4450 \\ -0.8020 \end{pmatrix},$$

and

$$\begin{pmatrix} \tilde{X}_{1(3)} \\ \tilde{X}_{2(3)} \\ \tilde{X}_{3(3)} \end{pmatrix} = \mu_3 \begin{pmatrix} X_{1(3)} \\ X_{2(3)} \\ X_{3(3)} \end{pmatrix} = \mu_3 \begin{pmatrix} 1 \\ -1.2470 \\ 0.5550 \end{pmatrix}.$$

For μ_1, one performs the matrix operation

$$\begin{pmatrix} \tilde{X}_{1(1)} \\ \tilde{X}_{2(1)} \\ \tilde{X}_{3(1)} \end{pmatrix}^T [M] \begin{pmatrix} \tilde{X}_{1(1)} \\ \tilde{X}_{2(1)} \\ \tilde{X}_{3(1)} \end{pmatrix} = \mu_1[1 \quad 1.8018 \quad 2.2468]m \begin{bmatrix} 1 & 0 & 0 \\ 0 & 1 & 0 \\ 0 & 0 & 1 \end{bmatrix} \mu_1 \begin{pmatrix} 1 \\ 1.8019 \\ 2.2468 \end{pmatrix} = 1$$

which gives

$$\mu_1^2 m\,(9.2946) = 1.$$

Therefore, $\mu_1 = \frac{0.3280}{\sqrt{m}}$. This leads to

$$\begin{pmatrix} \tilde{X}_{1(1)} \\ \tilde{X}_{2(1)} \\ \tilde{X}_{3(1)} \end{pmatrix} = \mu_1 \begin{pmatrix} 1 \\ 1.8019 \\ 2.2468 \end{pmatrix} = \frac{1}{\sqrt{m}} \begin{pmatrix} 0.3280 \\ 0.5910 \\ 0.7370 \end{pmatrix}.$$

Similarly, for μ_2, one performs the matrix operation

$$\begin{pmatrix} \tilde{X}_{1(2)} \\ \tilde{X}_{2(2)} \\ \tilde{X}_{3(2)} \end{pmatrix}^T [M] \begin{pmatrix} \tilde{X}_{1(2)} \\ \tilde{X}_{2(2)} \\ \tilde{X}_{3(2)} \end{pmatrix} = \mu_2^2 [1 \quad 0.4450 \quad -0.8020] m \begin{bmatrix} 1 & 0 & 0 \\ 0 & 1 & 0 \\ 0 & 0 & 1 \end{bmatrix} \begin{pmatrix} 1 \\ 0.4450 \\ -0.8020 \end{pmatrix} = 1$$

which gives

$$\mu_2^2 m\,(1.8412) = 1.$$

Therefore, $\mu_2 = \frac{0.7370}{\sqrt{m}}$. This results in

$$\begin{pmatrix} \tilde{X}_{1(2)} \\ \tilde{X}_{2(2)} \\ \tilde{X}_{3(2)} \end{pmatrix} = \frac{0.7370}{\sqrt{m}} \begin{pmatrix} 1 \\ 0.4450 \\ -0.8020 \end{pmatrix} = \frac{1}{\sqrt{m}} \begin{pmatrix} 0.7370 \\ 0.3280 \\ -0.5910 \end{pmatrix}.$$

For μ_3, one performs the matrix operation

$$\mu_3^2 [1 \quad -1.2470 \quad 0.5550] m \begin{bmatrix} 1 & 0 & 0 \\ 0 & 1 & 0 \\ 0 & 0 & 1 \end{bmatrix} \begin{pmatrix} 1 \\ -1.2470 \\ 0.5550 \end{pmatrix} = 1$$

which gives

$$\mu_3^2 m\,(2.8630) = 1.$$

Therefore, $\mu_3 = \frac{0.5910}{\sqrt{m}}$. This leads to

$$\begin{pmatrix} \tilde{X}_{1(3)} \\ \tilde{X}_{2(3)} \\ \tilde{X}_{3(3)} \end{pmatrix} = \mu_3 \begin{pmatrix} X_{1(3)} \\ X_{2(3)} \\ X_{3(3)} \end{pmatrix} = \frac{0.5910}{\sqrt{m}} \begin{pmatrix} 1 \\ -1.2470 \\ 0.5550 \end{pmatrix} = \frac{1}{\sqrt{m}} \begin{pmatrix} 0.5910 \\ -0.7370 \\ 0.3280 \end{pmatrix}.$$

From above, the normalized transformation matrix becomes

$$[\tilde{R}] = \frac{1}{\sqrt{m}}\left[\begin{pmatrix}0.3280\\0.5910\\0.7370\end{pmatrix} \quad \begin{pmatrix}0.7370\\0.3280\\-0.5910\end{pmatrix} \quad \begin{pmatrix}0.5910\\-0.7370\\0.3280\end{pmatrix}\right].$$

Applying this normalized transformation matrix to the matrix equation of motion,

$$[\tilde{R}]^T[M]\,[\tilde{R}](\ddot{\vec{q}}\,) + [\tilde{R}]^T[C\}[\tilde{R}](\dot{\vec{q}}) + [\tilde{R}]^T[K][\tilde{R}](\vec{q}) = [\tilde{R}]^T(\vec{F}) \quad (xiv)$$

so that the uncoupled equations of motion for the system are

$$\ddot{q}_r + 2\zeta_r\omega_r\dot{q}_r + \omega_r^2 q_r = f_r(t)\,, \qquad r = 1, 2, 3. \quad (xv)$$

Note that $[\tilde{R}]^T[M]\,[\tilde{R}] = [I]$, where $[I]$ is the unit or identity matrix of order 3, and operating $[\tilde{R}]^T[K][\tilde{R}]$ to give

$$[\tilde{R}]^T[K][\tilde{R}] = \frac{k}{m}\begin{bmatrix}0.4451^2 & 0 & 0\\ 0 & 1.2470^2 & 0\\ 0 & 0 & 1.8019^2\end{bmatrix} = \begin{bmatrix}\omega_1^2 & 0 & 0\\ 0 & \omega_2^2 & 0\\ 0 & 0 & \omega_3^2\end{bmatrix}.$$

The forcing term

$$[\tilde{R}]^T(\vec{F}) = \frac{1}{\sqrt{m}}\left[\begin{pmatrix}0.3280\\0.7370\\0.5910\end{pmatrix} \quad \begin{pmatrix}0.5910\\0.3280\\-0.7370\end{pmatrix} \quad \begin{pmatrix}0.7370\\-0.5910\\0.3280\end{pmatrix}\right](\vec{F})$$

$$= \frac{F\cos\omega t}{\sqrt{m}}\left[\begin{pmatrix}0.3280\\0.7370\\0.5910\end{pmatrix} \quad \begin{pmatrix}0.5910\\0.3280\\-0.7370\end{pmatrix} \quad \begin{pmatrix}0.7370\\-0.5910\\0.3280\end{pmatrix}\right]\begin{pmatrix}1\\1\\1\end{pmatrix}$$

$$[\tilde{R}]^T(\vec{F}) = \begin{pmatrix}f_1(t)\\f_2(t)\\f_3(t)\end{pmatrix} = \frac{F\cos\omega t}{\sqrt{m}}\begin{pmatrix}1.6560\\0.4740\\0.1820\end{pmatrix} = \begin{pmatrix}F_{10}\\F_{20}\\F_{30}\end{pmatrix}\cos\omega t\,,$$

$$\begin{pmatrix}F_{10}\\F_{20}\\F_{30}\end{pmatrix} = \frac{F}{\sqrt{m}}\begin{pmatrix}1.6560\\0.4740\\0.1820\end{pmatrix}.$$

Now, the steady-state solution of Equation (xv) can be obtained as that for a single dof system in Chapter 3. Thus,

$$q_r = Q_r \cos(\omega\, t - \varphi_r) \qquad (xvi)$$

where

$$Q_r = \left(\frac{F_{r0}}{\omega_r^2}\right)\frac{1}{\sqrt{\left[1-\left(\frac{\omega}{\omega_r}\right)^2\right]^2 + \left(2\zeta_r \frac{\omega}{\omega_r}\right)^2}} \qquad (xviia)$$

and

$$\varphi_r = tan^{-1}\left[\frac{2\zeta_r \dfrac{\omega}{\omega_r}}{1-\left(\frac{\omega}{\omega_r}\right)^2}\right]. \qquad (xviib)$$

By substituting the given data into Equation $(xvii)$, one obtains

$$q_1 = Q_1 \cos(\omega\, t - \varphi_1) = \left(\frac{F_{10}}{\omega_1^2}\right)\frac{\cos(\omega\, t - \varphi_1)}{\sqrt{\left[1-\left(\frac{\omega}{\omega_1}\right)^2\right]^2 + \left(2\zeta_1 \frac{\omega}{\omega_1}\right)^2}}.$$

But

$$\left(\frac{\omega}{\omega_1}\right)^2 = \frac{\left(\frac{7}{4}\sqrt{\frac{k}{m}}\right)^2}{0.1981\frac{k}{m}} = 15.4594,\ \left[1-\left(\frac{\omega}{\omega_1}\right)^2\right]^2 = (-14.4594)^2 = 209.0743,$$

$$\left(2\zeta_1 \frac{\omega}{\omega_1}\right)^2 = [2(0.01)]^2\ (15.4594) = 0.0062,\ \ 2\zeta_1 \frac{\omega}{\omega_1} = 0.0786,$$

$$\varphi_1 = tan^{-1}\left[\frac{2\zeta_1 \dfrac{\omega}{\omega_1}}{1-\left(\frac{\omega}{\omega_1}\right)^2}\right] = tan^{-1}\left[\frac{0.0786}{-14.4594}\right] = tan^{-1}(-\,0.0054)$$

and therefore,

$$q_1 = \left(\frac{0.5781\, F\sqrt{m}}{k}\right) cos\left(1.75\sqrt{\frac{k}{m}}t - \varphi_1\right). \qquad (xviii)$$

Similarly,

$$q_2 = \left(\frac{0.3143\ F\sqrt{m}}{k}\right) cos\left(1.75\sqrt{\frac{k}{m}}t - \varphi_2\right), \qquad (xix)$$

where $\varphi_2 = tan^{-1}(-\,0.0292)$, and

$$q_3 = \left(\frac{0.9249\ F\sqrt{m}}{k}\right) cos\left(1.75\sqrt{\frac{k}{m}}t - \varphi_3\right), \qquad (xx)$$

where $\varphi_3 = tan^{-1}(\,0.3415)$.

Finally, the steady-state responses in the original coordinates are given by

$$(\vec{x}) = [\tilde{R}](\vec{q}) = \frac{1}{\sqrt{m}}\left[\begin{pmatrix}0.3280\\0.5910\\0.7370\end{pmatrix} \begin{pmatrix}0.7370\\0.3280\\-\,0.5910\end{pmatrix} \begin{pmatrix}0.5910\\-\,0.7370\\0.3280\end{pmatrix}\right]\begin{pmatrix}q_1\\q_2\\q_3\end{pmatrix}.$$

Substituting for Equations $(xviii), (xix)$, and (xx), one obtains

$$x_1 = \left(\frac{0.1896\ F}{k}\right) cos\left(1.75\sqrt{\frac{k}{m}}t - \varphi_1\right) + \left(\frac{0.2316\ F}{k}\right) cos\left(1.75\sqrt{\frac{k}{m}}t - \varphi_2\right) + \left(\frac{0.5466\ F}{k}\right) cos\left(1.75\sqrt{\frac{k}{m}}t - \varphi_3\right), \qquad (xxi)$$

$$x_2 = \left(\frac{0.3417\ F}{k}\right) cos\left(1.75\sqrt{\frac{k}{m}}t - \varphi_1\right) + \left(\frac{0.1031\ F}{k}\right) cos\left(1.75\sqrt{\frac{k}{m}}t - \varphi_2\right) - \left(\frac{0.6817\ F}{k}\right) cos\left(1.75\sqrt{\frac{k}{m}}t - \varphi_3\right), \qquad (xxii)$$

$$x_3 = \left(\frac{0.4261\,F}{k}\right) cos\left(1.75\sqrt{\frac{k}{m}}t - \varphi_1\right)$$
$$-\left(\frac{0.1858\,F}{k}\right) cos\left(1.75\sqrt{\frac{k}{m}}t - \varphi_2\right)$$
$$+\left(\frac{0.3034\,F}{k}\right) cos\left(1.75\sqrt{\frac{k}{m}}t - \varphi_3\right). \qquad (xxiii)$$

5.6 Responses of Forced Mdof Systems with Non-Proportional Damping: Mechanical Impedance Method

The mechanical impedance (or simply impedance) method is applicable to systems with mdof if the excitations are harmonic or periodic. However, the amount of algebraic manipulation increases significantly as the number of dof of the system is increased. In what follows a 3dof system is considered to further illustrate the steps and amount of algebraic manipulation.

Example 5.6

Consider the three-dof system shown in Figure 5.9.

(a) From first principle, derive the equations of motion for the system shown in Figure 5.9 in which $f_1 = f_1(t) = F_1 \cos \omega t$.

(b) By applying the *mechanical impedance method*, evaluate the steady state responses of the system in Figure 5.9 in which the masses, spring, and damping constants are respectively, $m_1 = m_2 = m_3 = 1$ kg, $k_1 = k_2 = k_3 = 1\ \text{N/s}$, and $c_1 = c_2 = c_3 = 1\ \text{N.s/m}$. The applied frequency $\omega = 1\ \text{rad/s}$.

Solution:

(a) ***Equation of motion***

The equations of motion can be obtained by applying Newton's second law of motion, $\sum F = ma$ to every mass. By making use of the FBD shown in Figure 5.9a, one has

$$m_1\ddot{x}_1 = -k_1x_1 - k_2(x_1 - x_2) - c_1\dot{x}_1 - c_2(\dot{x}_1 - \dot{x}_2),$$

$$m_2\ddot{x}_2 = k_2(x_1 - x_2) - k_3(x_2 - x_3) + c_2(\dot{x}_1 - \dot{x}_2) - c_3(\dot{x}_2 - \dot{x}_3),$$

$$m_3\ddot{x}_3 = k_3(x_2 - x_3) + c_3(\dot{x}_2 - \dot{x}_3).$$

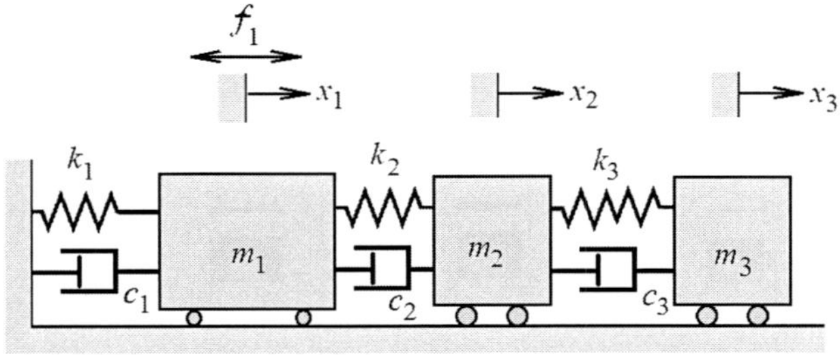

Figure 5.9: A system of 3dof under periodic excitation.

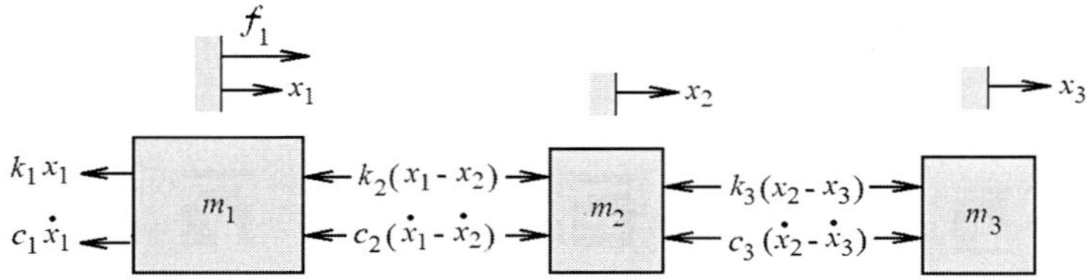

Figure 5.9a: FBD of system in Figure 5.9.

Rearranging terms,

$$m_1\ddot{x}_1 + (k_1 + k_2)x_1 - k_2x_2 + (c_1 + c_2)\dot{x}_1 - c_2\dot{x}_2 = F_1 cos\, \omega t,$$

$$m_2\ddot{x}_2 - k_2x_1 + (k_2 + k_3)x_2 - k_3x_3 - c_2\dot{x}_1 + (c_2 + c_3)\dot{x}_2 - c_3\dot{x}_3 = 0,$$

$$m_3\ddot{x}_3 + k_3x_3 - k_3x_2 + c_3\dot{x}_3 - c_3\dot{x}_2 = 0\,.$$

Expressing in matrix form,

$$\begin{bmatrix} m_1 & 0 & 0 \\ 0 & m_2 & 0 \\ 0 & 0 & m_3 \end{bmatrix} \begin{pmatrix} \ddot{x}_1 \\ \ddot{x}_2 \\ \ddot{x}_3 \end{pmatrix} + \begin{bmatrix} (c_1 + c_2) & -c_2 & 0 \\ -c_2 & (c_2 + c_3) & -c_3 \\ 0 & -c_3 & c_3 \end{bmatrix} \begin{pmatrix} \dot{x}_1 \\ \dot{x}_2 \\ \dot{x}_3 \end{pmatrix}$$

$$+\begin{bmatrix}(k_1+k_2) & -k_2 & 0\\ -k_2 & (k_2+k_3) & -k_3\\ 0 & -k_3 & k_3\end{bmatrix}\begin{pmatrix}x_1\\x_2\\x_3\end{pmatrix}=\begin{pmatrix}F_1\\0\\0\end{pmatrix}cos\,\omega t. \quad (i)$$

Equation (i) is the required equation of motion for the system.

(b) ***Responses by mechanical impedance method***

In the *mechanical impedance method* and for convenience, one can write $cos\,\omega t$ as $Re\,\{e^{i\omega t}\}$ such that $x_1 = X_i\,cos\,\omega t,\quad i = 1,2,3$ can be written as $x_i = Re\{X_i e^{i\omega t}\}$, where $Re\{\,.\,\}$ denotes the real part of the enclosing complex variable. Without loss of generality, one can disregard $Re\{\,.\,\}$ for the time being so that the displacements can be written as $x_i = X_i e^{i\omega t}$. Then, the velocity and acceleration are given by, respectively, $\dot{x}_i = i\omega X_i e^{i\omega t}$, $\ddot{x}_i = -\omega^2 X_i e^{i\omega t}$.

Substituting the above expressions into Eq. (i) and upon cancelling $e^{i\omega t}$ on both sides of the resulting matrix equation of motion, one has

$$\begin{bmatrix}-\omega^2 m_1 & 0 & 0\\ 0 & -\omega^2 m_2 & 0\\ 0 & 0 & -\omega^2 m_3\end{bmatrix}\begin{pmatrix}X_1\\X_2\\X_3\end{pmatrix}$$

$$+\begin{bmatrix}i\omega(c_1+c_2) & -i\omega c_2 & 0\\ -i\omega c_2 & i\omega(c_2+c_3) & -i\omega c_3\\ 0 & -i\omega c_3 & i\omega c_3\end{bmatrix}\begin{pmatrix}X_1\\X_2\\X_3\end{pmatrix}$$

$$+\begin{bmatrix}(k_1+k_2) & -k_2 & 0\\ -k_2 & (k_2+k_3) & -k_3\\ 0 & -k_3 & k_3\end{bmatrix}\begin{pmatrix}X_1\\X_2\\X_3\end{pmatrix}=\begin{pmatrix}F_1\\0\\0\end{pmatrix}.$$

Simplifying,

$$[D_{ij}]\begin{pmatrix}X_1\\X_2\\X_3\end{pmatrix}=\begin{pmatrix}F_1\\0\\0\end{pmatrix} \quad (ii)$$

where the non-zero elements or entries of the coefficient matrix $[D_{ij}]$ are given as

$$D_{11} = (k_1+k_2) + i\omega(c_1+c_2) - \omega^2 m_1, \qquad D_{12} = D_{21} = -k_2 - i\omega c_2,$$

$$D_{22} = (k_2+k_3) + i\omega(c_2+c_3) - \omega^2 m_2, \qquad D_{23} = D_{32} = -k_3 - i\omega c_3,$$

$$D_{33} = k_3 - \omega^2 m_3 + i\omega c_3.$$

Substituting for the given data, one obtains

$$\begin{bmatrix} 1+i2 & -1-i & 0 \\ -1-i & 1+i2 & -1-i \\ 0 & -1-i & i \end{bmatrix} \begin{pmatrix} X_1 \\ X_2 \\ X_3 \end{pmatrix} = \begin{pmatrix} F_1 \\ 0 \\ 0 \end{pmatrix}.$$

Solving this matrix equation by Cramer's rule,

$$X_1 = \frac{\begin{vmatrix} F_1 & -1-i & 0 \\ 0 & 1+i2 & -1-i \\ 0 & -1-i & i \end{vmatrix}}{\begin{vmatrix} 1+i2 & -1-i & 0 \\ -1-i & 1+i2 & -1-i \\ 0 & -1-i & i \end{vmatrix}},$$

$$X_2 = \frac{\begin{vmatrix} 1+i2 & F_1 & 0 \\ -1-i & 0 & -1-i \\ 0 & 0 & i \end{vmatrix}}{\begin{vmatrix} 1+i2 & -1-i & 0 \\ -1-i & 1+i2 & -1-i \\ 0 & -1-i & i \end{vmatrix}},$$

$$X_3 = \frac{\begin{vmatrix} 1+i2 & -1-i & F_1 \\ -1-i & 1+i2 & 0 \\ 0 & -1-i & 0 \end{vmatrix}}{\begin{vmatrix} 1+i2 & -1-i & 0 \\ -1-i & 1+i2 & -1-i \\ 0 & -1-i & i \end{vmatrix}}.$$

Operating on the above determinants, one has

$$X_1 = \frac{-F_1(2+i)}{2-i5}, \qquad X_2 = \frac{-F_1(1-i)}{2-i5}, \qquad X_3 = \frac{2F_1(i)}{2-i5}.$$

Making rationalization, one has

$$X_1 = \frac{-F_1(2+i)(2+i5)}{(2-i5)(2+i5)} = \frac{-F_1(-1+i12)}{29} = \frac{F_1(1-i12)}{29}$$

From the theory of complex variables,

$$X_1 = \frac{F_1(1 - i12)}{29} = X_{10}e^{i\varphi_1},$$

where

$$X_{10} = \frac{F_1}{29}\sqrt{1^2 + 12^2}\ , \qquad \varphi_1 = tan^{-1}(-12) = -85.2364^o.$$

Therefore,

$$X_{10} = 0.4152\ F_1, \qquad \varphi_1 = -85.2364^o\ .$$

Thus,

$$x_1 = Re\{X_1e^{i\omega t}\} = Re\{X_{10}e^{i(\omega t + \varphi_1)}\}$$

$$= 0.4152\ F_1 cos(\omega t - 85.2364^o). \qquad (iii)$$

Similarly,

$$X_2 = \frac{-F_1(1 - i)(2 + i5)}{(2 - i5)(2 + i5)} = \frac{-F_1(7 + i3)}{29} = \frac{F_1(-7 - i3)}{29}$$

From the theory of complex variables,

$$X_2 = \frac{F_1(-7 - i3)}{29} = X_{20}e^{i\varphi_2},$$

where

$$X_{20} = \frac{F_1}{29}\sqrt{7^2 + 3^2}\ , \qquad \varphi_2 = tan^{-1}\left(\frac{3}{7}\right) = 23.2^o.$$

Therefore,

$$X_{20} = 0.2626\ F_1, \qquad \varphi_2 = 23.2^o\ ,$$

$$x_2 = Re\{X_2e^{i\omega t}\} = Re\{X_{20}e^{i(\omega t + \varphi_2)}\}$$

$$= 0.2626\ F_1 cos(\omega t + 23.2^o). \qquad (iv)$$

Finally,

$$X_3 = \frac{2\ F_1\ i(2 + i5)}{(2 - i5)(2 + i5)} = \frac{2\ F_1(-5 + i2)}{29}.$$

From the theory of complex variables,

$$X_3 = \frac{2\,F_1(-\,5 + i2)}{29} = \,X_{30}e^{i\varphi_3},$$

Where

$$X_{30} = \frac{2\,F_1}{29}\,\sqrt{(-\,5)^2 + 2^2}\ , \qquad \varphi_3 = \,tan^{-1}\left(-\,\frac{2}{5}\right) = \,-\,21.8014^o.$$

Therefore,

$$X_{30} = 0.3714\,F_1, \qquad \varphi_3 = \,-\,21.8014^o\,,$$

$$x_3 = Re\{X_3e^{i\omega t}\} = Re\{X_{30}e^{i(\omega t\,+\,\varphi_3)}\}$$

$$= 0.3714\,F_1cos(\omega t - 21.8014^o). \qquad (v)$$

Equations (iii) through (v) give the responses of the system.

5.7 Responses of Forced Mdof Systems with Non-Proportional Damping: Space Vector Approach

As demonstrated in Section 4.7 even for a tdof system the amount of algebraic manipulation is very substantial. Therefore, when solutions of mdof systems are required one often resorts to applying some computer software. The application of a particular computer software for mdof systems is not pursued here as it is beyond the scope of the present book.

However, for completeness and to provide detailed steps of the space vector approach the following linear *n*dof system is considered. The matrix equation of motion for this system is given by

$$[M](\ddot{\vec{x}}\,) + [C\}(\dot{\vec{x}}\,) + \,[K](\vec{x}\,) = (\vec{F}\,) \qquad (5.44)$$

where the mass, and stiffness matrices have been defined previously in this chapter while the non-proportional damping is general, and the displacement as well as forcing vector are given, respectively by

$$(\vec{x}\,) = \begin{pmatrix} x_1 \\ x_2 \\ \vdots \\ x_n \end{pmatrix}, \quad (\vec{F}\,) = \begin{pmatrix} f_1 \\ f_2 \\ \vdots \\ f_n \end{pmatrix}$$

in which f_i are applied forces.

In the space vector approach the n-second order differential equations of motion are transformed into $2n$-first order differential equations. To this end, one writes

$$(\vec{Y}) = \begin{pmatrix} \vec{x} \\ \dot{\vec{x}} \end{pmatrix}, \quad (\vec{x}) = \begin{pmatrix} x_1 \\ x_2 \\ \vdots \\ x_n \end{pmatrix}, \quad (\dot{\vec{x}}) = \begin{pmatrix} \dot{x}_1 \\ \dot{x}_2 \\ \vdots \\ \dot{x}_n \end{pmatrix},$$

so that Equation (5.44) can be rewritten as

$$(\dot{\vec{Y}}) = \begin{bmatrix} [0] & [I] \\ -[M]^{-1}[K] & -[M]^{-1}[C] \end{bmatrix} (\vec{Y}) + \begin{pmatrix} \vec{0} \\ \vec{F} \end{pmatrix} \qquad (5.45)$$

where $(\vec{Y})$ is a $2n \times 1$ vector of generalized displacements.

Step 1: *Solution of eigenvalues and eigenvectors*

In order to evaluate the eigenvalues and eigenvector, the *coefficient* or *amplification matrix* is required.

Let the coefficient matrix be

$$[A] = \begin{bmatrix} [0] & [I] \\ -[M]^{-1}[K] & -[M]^{-1}[C] \end{bmatrix},$$

where $[I]$ is the $n \times 1$ *unit* or *identity matrix*. The eigenvalues of $[A]$ are obtained from the frequency equation, $|\lambda[I] - [A]| = 0$ or $|[A] - \lambda[I]| = 0$. Solving this frequency equation, one obtains the eigenvalues $\lambda_1, \lambda_2, \ldots, \lambda_{2n}$.

For the eigenvectors or normal modes or mode shapes, one makes use of the following equation

$$([A] - \lambda[I])(\vec{Y}) = \begin{pmatrix} \vec{0} \\ \vec{0} \end{pmatrix}. \qquad (5.46)$$

Substituting λ_1 into Equation (5.46), one obtains the corresponding eigenvector $(\vec{Y})_{(1)}$ where the subscript (1) denotes the eigenvector associated with λ_1.

Similarly, substituting the remaining eigenvalues, one at a time, the corresponding eigenvectors can be determined.

Step 2: ***Diagonalization of amplification matrix***

In order to solve for $(\vec{Y})$ the amplification matrix has to be uncoupled. To this end, the state vector $(\vec{Y})$ has to be transformed into another vector. Let the latter be related to

$$(\vec{Y}) = [R](\vec{Z}) \tag{5.47}$$

where $(\vec{Z}) = \begin{pmatrix} z_1 \\ z_2 \\ \vdots \\ z_{2n} \end{pmatrix}$ and the transformation matrix $[R]$ whose columns are the eigenvectors obtained in the foregoing. That is,

$$[R] = \left[(\vec{Y})_{(1)} \quad (\vec{Y})_{(2)} \quad \cdots \quad (\vec{Y})_{(2n)} \right]. \tag{5.48}$$

Substituting the transformation into Eq. (5.45), one obtains

$$[R]\begin{pmatrix} \dot{z}_1 \\ \dot{z}_2 \\ \vdots \\ \dot{z}_{2n} \end{pmatrix} = [A][R]\begin{pmatrix} z_1 \\ z_2 \\ \vdots \\ z_{2n} \end{pmatrix} + \begin{pmatrix} \vec{0} \\ \vec{F} \end{pmatrix}. \tag{5.49}$$

Pre-multiplying the inverse of $[R]$, it becomes

$$\begin{pmatrix} \dot{z}_1 \\ \dot{z}_2 \\ \vdots \\ \dot{z}_{2n} \end{pmatrix} = [R]^{-1}[A][R]\begin{pmatrix} z_1 \\ z_2 \\ \vdots \\ z_{2n} \end{pmatrix} + [R]^{-1}\begin{pmatrix} \vec{0} \\ \vec{F} \end{pmatrix}, \tag{5.50}$$

where $[R]^{-1}[A][R] = [D]$ is a diagonal matrix such that Eq. (5.50) is uncoupled and therefore z_r, $r = 1, 2, \cdots, 2n$ can be determined individually as a single first order differential equation. It should be noted that the diagonal elements of $[D]$ are the eigenvalues of $[A]$. This fact was shown for a tdof system that was presented in Section 4.7.

Step 3: ***Forced responses***

The solution of the uncoupled equations in Equation (5.50) is similar to that for a tdof system in Section 4.7. Once Equation (5.50) is solved the state vector $(\vec{Y}) = [R](\vec{Z})$ can simply be determined.

References

Amirouche, Farid M. L. 1992. Computational Methods in Multibody Dynamics. Englewood Cliffs, New Jersey: Prentice-Hall.
Clough, R. W. and Penzien, J. 1975. Dynamics of Structures. New York: McGraw-Hill.
Meirovitch, L. 1997. Principles and Techniques of Vibration. Englewood Cliffs, New Jersey: Prentice-Hall.
Rao, S. S. 1995. Mechanical Vibrations, Third Edition. New York: McGraw-Hill.

Exercise Questions

Q13. State the kinetic energy, potential energy and Lagrangian of the system in Example 5.3. Hence, derive the equations of motion.

Q14. The figure shows the triple pendulum which contains 3 point masses connected by two light strings.

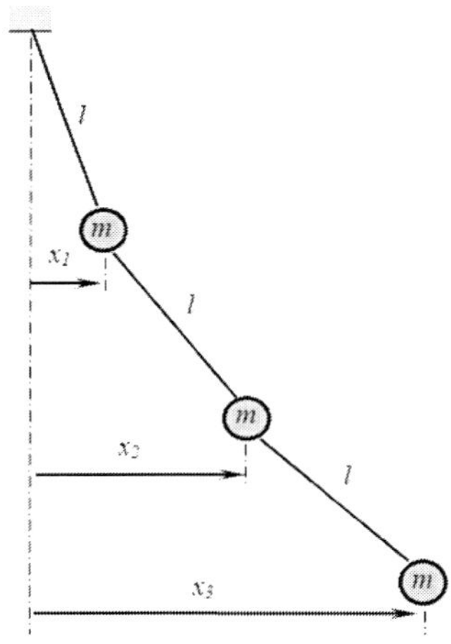

Figure 5Q14 : A triple pendulum.

(a) Assuming small amplitude vibration, derive the equations of motion for the triple pendulum by the Newtonian method.

(b) Repeat (a) by using the Lagrangian method.

(c) Hence, show that eigenfrequencies of the system are $0.6448\sqrt{g/l}$, $1.5147\sqrt{g/l}$, $2.5080\sqrt{g/l}$.

Q15. The figure shows a rigid two-dimensional platform of dimension $4l \times 2l$. The platform is supported by four springs of the same spring constant k. A machinery producing an out-of-balance force F rotating at an angular speed ω is mounted on top of the platform. At time t, F is inclined at ωt to the horizontal. For both the platform and the machinery, the geometric centre and CG (the centre of gravity) are coincident. For the platform, its mass and the moment of inertia about its CG are $10m$ and $15ml^2$. Those for the machinery are m and $ml^2/5$.

(a) If $G(x_G, y_G)$ is the overall CG of the platform and the machinery, determine x_G and y_G.

(b) Determine the moment of inertia I_G of the platform and the machinery about G.

(c) Assuming small amplitude vibration, express the spring tensions in terms of x_G, y_G, the horizontal displacement u of G, the vertical displacement v of G and the rotation θ.

(d) With u, v and θ taken to be the dofs, derive the equation of motion for the vibrating system in terms of I_G, x_G and y_G.

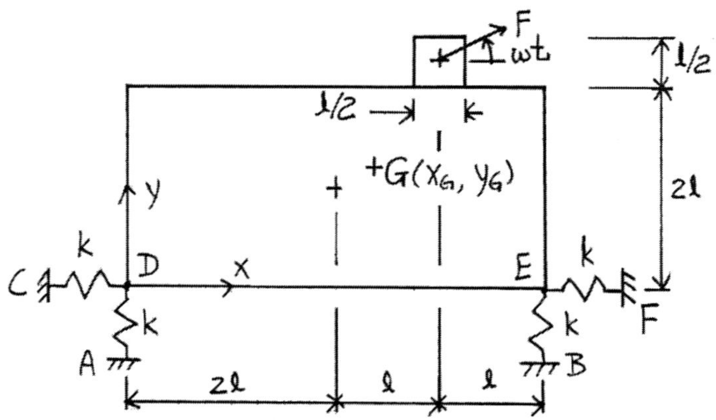

Figure 5Q15: A platform with rotatory machinery mounted.

Q16. The equation of motion for a vibration system is

$$[M]\{\ddot{D}\} + \frac{1}{10}[M]\{\dot{D}\} + [K]\{D\} = \{F\}$$

where

$$[M] = \begin{bmatrix} 2 & 1 & 0 \\ 1 & 2 & -1 \\ 0 & -1 & 1 \end{bmatrix}, \quad [K] = \begin{bmatrix} 8 & -2 & 1 \\ -2 & 4 & -2 \\ 1 & -2 & 8 \end{bmatrix},$$

$$\{F\} = \begin{Bmatrix} \sin 2t \\ 0 \\ 0 \end{Bmatrix}$$ and $\{D\}$ the vector of dofs.

(a) Given that the eigenvectors of systems are

$$\begin{Bmatrix} -0.3471 \\ 0.5732 \\ 0.7422 \end{Bmatrix}, \begin{Bmatrix} 0.7024 \\ -0.3485 \\ 0.6206 \end{Bmatrix} \text{ and } \begin{Bmatrix} 0.4739 \\ 0.8796 \\ 0.0418 \end{Bmatrix},$$

derive the normal modes and the eigenfrequencies of the system.

(b) State that the modal matrix $[R]$.

(c) By introducing the vector of modal coordinates $\{\alpha\} = \{\alpha_1, \alpha_2, \alpha_3\}^T$ through $\{D\} = [R]\{\alpha\}$ and ignoring the small non-zero off-diagonal entries in $[R]^T[M]\,[R]$ and $[R]^T[K]\,[R]$, decouple the system equation.

Q17. Solve the steady-state response of the vibration system in the last question by the mechanical impedance method.

CHAPTER 6

VIBRATION OF CONTINUOUS SYSTEMS: STRINGS, RODS, AND BEAMS

Continuous systems are those that continuously distribute mass and elasticity. These systems are known as *distributed parameter* models. Strings, rods and beams are typical continuous systems in real structures. A continuous system can be imaged as a combination of infinitely many infinitesimal mass elements and therefore possesses an infinite number of degrees of freedom. Stiffness, damping and other physical parameters of a continuous system as well as response are functions of time and spatial coordinates.

As stated in the former chapters, the discrete system has a finite number of degrees of freedom, and the vibrations of discrete systems are governed by ordinary differential equations. However, the vibrations of continuous systems are governed by partial differential equations with both temporal and spatial independent variables. Nevertheless, because the discrete and continuous systems represent merely two mathematical models of vibrating systems, they display similar dynamical behaviour in the physical essence, and the mathematical formulation for a given continuous system can be derived as a limit case of that of a discrete system.

In this chapter, various free and forced vibrations are discussed for some continuous systems, mainly beams.

Basic assumptions are made as follows: the elastic body or system is *homogeneous* and *isotropic*, and it obeys *Hooke's law* within the elastic limit. Hence, linear dynamic system analysis is considered here, and the vibration analysis can be realized by exact solutions in terms of natural modes.

6.1 Vibrating Cables or Strings and Rods

In this section the vibration of cables or strings, axial or longitudinal vibration of rods, and torsional vibration of rods are presented.

6.1.1 Vibrating cables or strings

A flexible cable or string of mass per unit length μ is stretched under tension T (not to be confused with the symbol for kinetic energy).

- Assume that the lateral deflection y (strictly, this is not the coordinate y it is usually represented as v in more advanced books of vibration) of the string to be small such that the change in tension T with deflection is negligible.

Consider the free-body diagram (FBD) of an elementary length dx of the string as shown in Figure 6.1. The slope or angular displacement θ is a function of x and therefore at point A it is $\theta(x)$ whereas at point B it is $\theta(x + dx)$. Note that strictly speaking, $\theta(x + dx)$ should read as $\theta(x + dx,t)$ since the angular displacement is a function of space and time. But for conciseness the time variable t has been disregarded in the argument of θ henceforth, unless it is required for clarity.

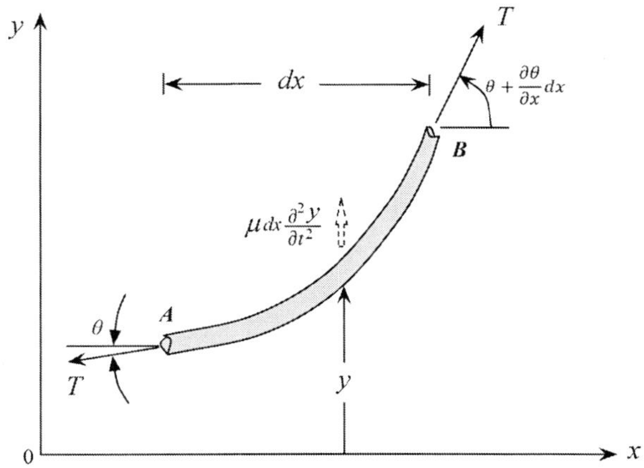

Figure 6.1: Elementary length of vibrating string.

By Taylor expansion,

$$\theta(x + dx) = \theta(x) + \frac{\partial\theta}{\partial x}dx + hot$$

where *hot* denotes the higher order terms which will be disregarded for linear dynamic system analysis.

By Newton's second law, the equation of motion for the element shown in Figure 6.1 becomes

$$\mu\, dx \frac{\partial^2 y}{\partial t^2} = T\, sin[\theta(x + dx)] \; - \; T\, sin\theta.$$

Application of the above result of the Taylor expansion, it leads to

$$\mu\, dx \frac{\partial^2 y}{\partial t^2} = T\, sin\left(\theta + \frac{\partial \theta}{\partial x} dx\right) \; - \; T\, sin\theta.$$

Since θ is small, one has

$$sin\theta \approx \theta, \quad sin\left(\theta + \frac{\partial \theta}{\partial x} dx\right) \approx \theta + \frac{\partial \theta}{\partial x} dx,$$

so that the equation of motion becomes

$$\mu\, dx \frac{\partial^2 y}{\partial t^2} = T\left(\theta + \frac{\partial \theta}{\partial x} dx\right) - T\theta \quad \text{or} \quad \frac{\partial^2 y}{\partial t^2} = \left(\frac{T}{\mu}\right)\frac{\partial \theta}{\partial x}.$$

Since the slope of the string is $= \frac{\partial y}{\partial x}$, the last equation can be written as

$$\frac{\partial^2 y}{\partial t^2} = c^2 \frac{\partial^2 y}{\partial x^2} \tag{6.1}$$

where $c = \sqrt{T/\mu}$ is known as the *velocity of wave propagation* of the string. In order to solve for this linear second order partial differential equation (p.d.e.) the method *separable variables* is employed. In this latter method, the solution is assumed to be a product of two single variable function

$$y(x,t) = \; Y(x)\, U(t). \tag{6.2}$$

By substituting into Eq. (6.1) and writing $Y(t)$ and $U(t)$ simply as Y and U, and upon dividing both sides of the resulting equation by YU, one obtains

$$\frac{1}{U}\frac{d^2 U}{dt^2} = \frac{c^2}{Y}\frac{d^2 Y}{dx^2}\,. \tag{6.3}$$

Since the lhs of this equation is independent of x, whereas the rhs is independent of t, it follows that each side must be a constant. Letting this constant be $-\omega^2$ so that Equation (6.3) becomes

$$\frac{1}{U}\frac{d^2U}{dt^2} = \frac{c^2}{Y}\frac{d^2Y}{dx^2} = -\omega^2 .$$

This gives two ordinary differential equations (o.d.e.) as

$$\frac{d^2U}{dt^2} + \omega^2 U = 0, \quad \text{and} \quad \frac{d^2Y}{dx^2} + \left(\frac{\omega}{c}\right)^2 Y = 0 .$$

The solution to the second o.d.e. is given by

$$Y = A\sin\left(\frac{\omega}{c}x\right) + B\cos\left(\frac{\omega}{c}x\right), \tag{6.4a}$$

whereas the solution to the first o.d.e. is

$$U = C\, sin\, \omega t + D cos\omega t, \tag{6.4b}$$

where the arbitrary constants A, B, C and D depend on the boundary and initial conditions of the system.

Example 6.1

If the string is stretched between two fixed points with distance L between them, the boundary conditions are: $y(0,t) = y(L,t) = 0$. Determine the general solution $y(x,t)$ of the stretched string.

Solution:

With reference to Equation (6.4) and the boundary condition that $y(0,t) = 0$, one requires that $B = 0$ so that the solution becomes

$$Y = A\sin\left(\frac{\omega}{c}x\right)(C\, sin\, \omega t + D cos\omega t) \tag{i}$$

The boundary condition that $y(L,t) = 0$ leads to the equation

$$sin\left(\frac{\omega}{c}L\right) = 0 \text{ which gives } \frac{\omega_n L}{c} = n\pi, \quad \text{with } n = 1, 2, 3, \dots, \infty.$$

By definition, $c = f\lambda$ where *f* is the frequency of oscillation and λ is the wavelength. One can then write

$$\frac{\omega_n}{c} = \frac{n\pi}{L}. \qquad (ii)$$

Therefore, by making use of Equation (6.4*a*) and recalling that $B = 0$, the *mode shape* or *normal mode* or *eigenvector* of the vibrating string is given as

$$Y = A\, sin\left(n\pi \frac{x}{L}\right). \qquad (iii)$$

Note that this equation is originated from the rhs of Equation (*i*).

In general, free vibration of a string contains many of the normal modes and the equation for the displacement may be written as

$$y(x,t) = \sum_{n=1}^{\infty} \sin\left(n\pi\frac{x}{L}\right)(C_n\, sin\, \omega_n t + D_n cos\omega_n t). \qquad (iv)$$

where the arbitrary constants C_n and D_n can be evaluated with the initial conditions of $y(x,0)$ and $dy(x,0)/dt$. The integer *n* is the number of mode shapes or eigenvectors.

6.1.2 Axial or longitudinal vibration of rods

Axial or longitudinal vibration of rods has many applications. For example, in reinforcing the strength of the foundation piling is frequently applied. In driving the pile into the ground with impacting hammer longitudinal vibration occurs.

- It is assumed that the rod is thin (meaning, the length to diameter of cross-section is 20 or higher), uniform cross-section area along the length, homogeneous, isotropic, and obeys Hooke's law.

Consider an element of length dx of the thin and uniform rod of cross-section area A, shown in Figure 6.2. Owing to the axial force P, the rod is undergoing deformation (displacement) $u(x,t)$ or simply written as $u(x)$ along the *x*-axis at x, and the displacement $u(x+dx,t)$ or simply as $u(x+dx)$ at $x+dx$.

By Taylor expansion,

$$u(x+dx) = u(x) + \frac{\partial u}{\partial x}dx + hot \quad \text{and}$$
$$P(x+dx) = P(x) + \frac{\partial P}{\partial x}dx + hot$$

where *hot* will be disregarded for linear dynamic system analysis as in the case of vibrating cable in Sub-section 6.1.1.

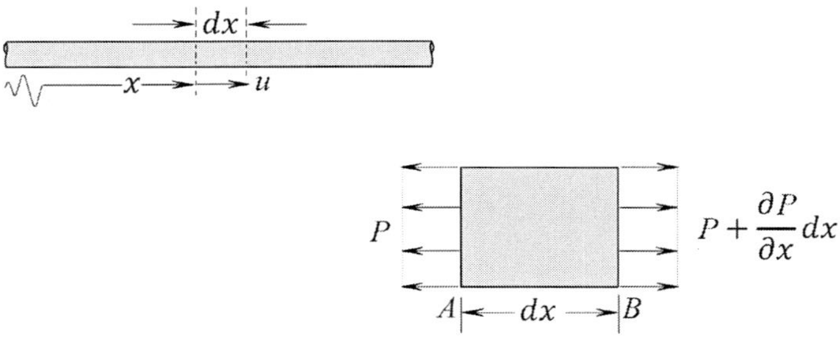

Figure 6.2: Elementary length of rod in longitudinal vibration.

By Hooke's law,

$$E = \frac{P}{A} \Big/ \frac{\partial u}{\partial x} \quad \text{or} \quad \frac{\partial u}{\partial x} = \frac{P}{AE}. \tag{6.5}$$

Partial differentiating Equation (6.5) w.r.t. x, one has

$$\frac{\partial^2 u}{\partial x^2} = \frac{1}{AE}\left(\frac{\partial P}{\partial x}\right). \tag{6.6}$$

By Newton's second law, the equation of motion for the element shown in Figure 6.2 becomes

$$(\rho A dx)\frac{\partial^2 u}{\partial t^2} = P(x+dx) - P(x).$$

Application of the above result of Taylor expansion for $P(x+dx)$, one obtains

$$(\rho A dx)\frac{\partial^2 u}{\partial t^2} = \left[P(x) + \frac{\partial P}{\partial x}dx\right] - P(x)$$

where ρ is the density of the rod (in mass per unit volume) and then $\rho A dx$ is the mass of the elementary length of the rod.

Simplifying, therefore, it becomes

$$(\rho A)\frac{\partial^2 u}{\partial t^2} = \frac{\partial P}{\partial x}. \tag{6.7}$$

Substituting this equation into Equation (6.6), one obtains

$$\frac{\partial^2 u}{\partial x^2} = \left(\frac{\rho}{E}\right)\frac{\partial^2 u}{\partial t^2} \quad \text{or} \quad \frac{\partial^2 u}{\partial t^2} = \left(\frac{E}{\rho}\right)\frac{\partial^2 u}{\partial x^2} \tag{6.8}$$

in which $c_L = \sqrt{\frac{E}{\rho}}$ is called the *longitudinal wave propagation of the rod.*

6.1.3 Torsional vibration of rods

Torsional vibration of rods is encountered in many engineering situations. For example, vibration in drill strings or rotating shafts or vibration in geared systems is of significant engineering interest.

- It is assumed that the rod is thin, uniform cross-section area along the length, homogeneous, isotropic, cross-sections of the rod initially perpendicular to the axis remain plane, and radii remain straight during deformation.
- Then, from the elementary elasticity theory the torque-twist relation is

$$\tau = GI_p\frac{\partial \theta}{\partial x} \tag{6.9}$$

where τ is the torque at section x, θ the twist or angular displacement about the x-axis at section x, G the shear modulus, and I_p the polar second moment of cross-section area.

Consider an element of length dx of the thin and uniform rod of cross-section area A, shown in Figure 6.3. Since the torque $\tau(x)$ is at section x and the torque at section $x + dx$ is $\tau(x + dx)$ one can apply Taylor expansion to give

$$\tau(x+dx) = \tau(x) + \frac{\partial \tau}{\partial x}dx + hot \qquad \text{and}$$

$$\tau(x+dx) - \tau(x) = \frac{\partial \tau}{\partial x}dx. \tag{6.10}$$

Now, consider the dynamic equilibrium of the element of the rod between x and $x + dx$,

$$\tau(x+dx) - \tau(x) = \frac{\partial \tau}{\partial x}dx = \rho I_p dx \frac{\partial^2 \theta}{\partial t^2} \tag{6.11}$$

where $\rho I_p dx$ is the *mass moment of inertia* of the element dx.

By partial differentiating Equation (6.9) w.r.t. x, one obtains

$$\frac{\partial \tau}{\partial x} = GI_p \frac{\partial^2 \theta}{\partial x^2}. \tag{6.12}$$

Substituting Equation (6.12) into (6.11), one has

$$\frac{\partial \tau}{\partial x}dx = GI_p \frac{\partial^2 \theta}{\partial x^2}dx = \rho I_p dx \frac{\partial^2 \theta}{\partial t^2}.$$

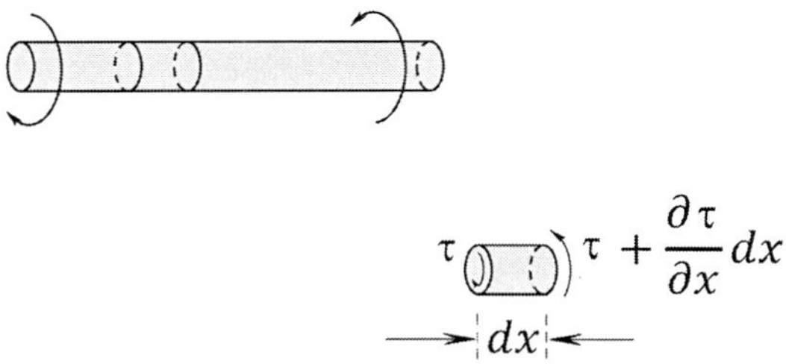

Figure 6.3: Elementary length of rod in torsional vibration.

Simplifying,

$$\left(\frac{G}{\rho}\right)\frac{\partial^2 \theta}{\partial x^2} = \frac{\partial^2 \theta}{\partial t^2} \qquad \text{or} \qquad \frac{\partial^2 \theta}{\partial t^2} = c_s^2 \frac{\partial^2 \theta}{\partial x^2} \tag{6.13}$$

in which $c_s = \sqrt{\frac{G}{\rho}}$ is the *velocity of propagation of shear waves* in the rod.

6.1.4 Vibration of moving taut strings

Vibration of a taut string that moves along its length has many practical applications. For example, tape in sound recording or moving cable car or belt transmission or wire production. An elementary length of the moving taut string is included in Figure 6.4. The symbols applied in Sub-section 6.1.1 are adopted here.

- Assume that the lateral deflection y of the string to be small such that the change in tension T with deflection is negligible. The question of how small is small in a moving string has been addressed by Mote. It was pointed out by him that the smallness defined for the vibration of a stationary string cannot be applied to the vibration of a moving string.

For linear analysis, as that considered in this sub-section, the definition of smallness can only apply to the case of high tension and low velocity. Even for the case of amplitude of vibration equal to a half percent of the length of the string, an amplitude-dependent tension term has to be included.

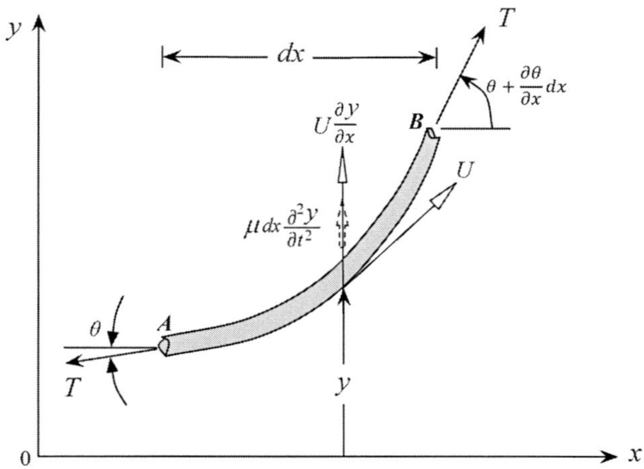

Figure 6.4: Elementary length of moving taut string.

With the above qualification of the definition of smallness in mind, one can consider the free-body diagram (FBD) of an elementary length dx of the string as shown in Figure 6.4. The taut string under a tension T moves with a constant speed U. An element of mass $\mu\, dx$ moves in the y-axis with velocity $v(x,t) = \frac{\partial y}{\partial t} + U\left(\frac{\partial y}{\partial x}\right)$. The slope or angular displacement θ is a function of x (strictly, it is a function of x and t, see, Sub-section 6.1.1) and therefore, following the steps in Sub-section 6.1.1, by application of Newton's law, one obtains

$$\mu\, dx \frac{dv}{dt} = T \frac{\partial \theta}{\partial x} dx\,. \tag{6.14}$$

Now, consider the term on the lhs of this equation. Since $v(x,t)$ is a function of space and time, therefore,

$$\frac{dv}{dt} = \frac{\partial v}{\partial x}\frac{\partial x}{\partial t} + \frac{\partial v}{\partial t} = \frac{\partial v}{\partial x} U + \frac{\partial v}{\partial t}.$$

Substituting this equation into Equation (6.14), one has

$$\mu\, dx \left(\frac{\partial v}{\partial x} U + \frac{\partial v}{\partial t}\right) = T \frac{\partial \theta}{\partial x} dx\,.$$

Substituting for $v(x,t) = \frac{\partial y}{\partial t} + U\left(\frac{\partial y}{\partial x}\right)$, the above equation becomes

$$\mu\, dx \left\{ \frac{\partial \left[\frac{\partial y}{\partial t} + U\left(\frac{\partial y}{\partial x}\right)\right]}{\partial x} U + \frac{\partial \left[\frac{\partial y}{\partial t} + U\left(\frac{\partial y}{\partial x}\right)\right]}{\partial t} \right\} = T \frac{\partial \theta}{\partial x} dx\,.$$

Upon simplification and using $\theta = \frac{\partial y}{\partial x}$, it becomes

$$(T - \mu U^2)\frac{\partial^2 y}{\partial x^2} - 2\mu U \frac{\partial^2 y}{\partial x \partial t} = \mu \frac{\partial^2 y}{\partial t^2}\,. \tag{6.15}$$

Upon dividing both sides by μ and recall $c = \sqrt{T/\mu}$ in Sub-section 6.1.1, one can re-writes the above equation as

$$(c^2 - U^2)\frac{\partial^2 y}{\partial x^2} - 2U \frac{\partial^2 y}{\partial x \partial t} = \frac{\partial^2 y}{\partial t^2}\,. \tag{6.16}$$

This equation becomes unstable when $U = c$. That is, when the speed of the string is equal to the speed of sound on the string the system is unstable.

6.1.5 Remarks

Equations (6.1), (6.8), and (6.13) are second order partial differential equations of the same form. Therefore, the solution steps for Equations (6.8) and (6.13) are similar to those for Equation (6.1).

6.2 Lateral Vibration of Beams

In this section lateral vibration of beams in many applications are considered. The most basic case is the Euler beam theory in transverse vibration. This is presented in Sub-section 6.2.1. Standard or simple and general Boundary conditions are dealt with in relatively detailed manner. Lateral vibration of beams with axial forces is considered in Sub-section 6.2.2. Lateral vibration of pipe conveying flowing fluid is provided in Sub-section 6.2.3 in which some representative computed natural frequencies and mode shapes are included. In Sub-section 6.2.4 lateral vibration of rotating beam is studied. Lateral vibration of beam with elastic foundation is included in Sub-section 6.2.5. The theory of Timoshenko beam is studied in Sub-section 6.2.6.

6.2.1 Euler beams

- The beam considered in this sub-section is assumed to be thin or slender and uniform along its length. By thin or slender, it means that the ratio of the length of the beam to the diameter of the beam is 20 or higher.
- The normal to the cross-section of the beam before deformation remains normal to the cross-section of the beam after deformation.
- It is assumed that the beam is homogeneous, isotropic, and obeys Hooke's law. Beams that satisfy the aforementioned assumptions are called *Euler beams*.

Consider an elementary length dx of the beam in Figure 6.5. The *sign convention* adopted in this sub-section is similar to that in (Thomson, 1993, 281, 306) (Seto, 1964, 137). It may be important to ask the question: *Why the directions of the shear forces are as they are in Figure 6.5*? To answer this question, it takes several forms.

- First, since the direction of the load intensity (distributed load) is pointing downward (see, Figure 6.5) and as a result the bending moments at both ends of the elementary length are sagging moments (that is, the moment at the lhs is clockwise and that on the rhs of the elementary length is counter-clockwise). Then if the elementary length is cut into several sections, the sum of the shear forces and bending moments on both sides of the section becomes zero. This is the basis of the free-body diagram (FBD).

- Second, the FBD is formed by recalling that the load intensity is pointing downward and when one cuts a section of the beam, on the rhs of the elementary length the shear force is $S + \frac{\partial s}{\partial x} dx$ is in the same direction as the load intensity $p(s)$. The shear force S on the lhs of the elementary length is pointing upward so that the shear forces on both sides of the elementary length give rise to the couples in the clockwise direction. These couples have to be balanced by the moments as shown in the figure in order to achieve equilibrium. Otherwise, the elementary length will rotate which is, of course, not true.

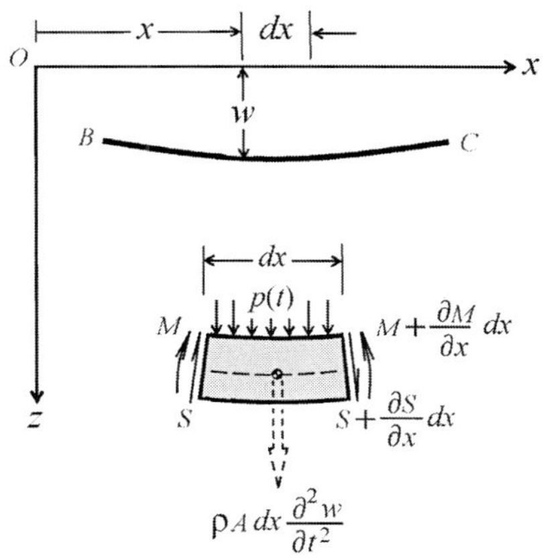

Figure 6.5: Elementary length of beam and FBD in lateral vibration.

- Third, for static equilibrium, slope of shear diagram at any section along the beam is negative of unit load at that point. This can be checked when the equation relating the slope of shear and load intensity is derived in the following.

Now, returning to the elementary length in Figure 6.5. If $M(x)$ or simply M is the bending moment about an axis at the lhs of the length, and bending moment about and axis at $x + dx$ is $M(x + dx)$ so that by applying the Taylor expansion one has

$$M(x + dx) = M(x) + \frac{\partial M}{\partial x} dx + hot.$$

Similarly, for the shear force, one has

$$S(x+dx) = S(x) + \frac{\partial S}{\partial x}dx + hot.$$

- Note that from this point onward one should recall that in Figure 6.5 counter-clockwise (ccw) is in fact clockwise (cw) in the right-hand screw rule. This is because the *z*-axis is pointing downward instead of pointing upward in the figure.
- The basis of equations derived from using the FBD is similar to that for a mass-spring-dashpot system.

Summing all forces vertically, one obtains

$$(\rho A dx)\frac{\partial^2 w}{\partial t^2} = \left(S + \frac{\partial S}{\partial x}dx\right) - S + p(x)dx$$

where $p(x)$ is the distributed load or load intensity which is expressed in force per unit length (in general, it is a function of space and time but for conciseness at this stage the time argument is disregarded here), w or $w(x,t)$ is the transverse or lateral deflection of the beam, and the remaining symbols have their usual meaning. Simplifying the last equation, one has

$$(\rho A)\frac{\partial^2 w}{\partial t^2} = \frac{\partial S}{\partial x} + p(x) \tag{6.17}$$

Now, summing all moments about an axis through the center of mass of the elementary length, one can obtain

$$M + S\left(\frac{dx}{2}\right) + \left(S + \frac{\partial S}{\partial x}dx\right)\left(\frac{dx}{2}\right) - \left[M(x) + \frac{\partial M}{\partial x}dx\right] = 0.$$

It may be appropriate to note that at this stage one should not cancel the common factor dx since the lhs contains the second order term $dx\,(dx)$. That is, one should cancel the second order term first such that the last equation becomes

$$S\,(dx) - \frac{\partial M}{\partial x}dx = 0\,.$$

Now, the common factor dx can be deleted. Thus,

$$\frac{\partial M}{\partial x} = S. \tag{6.18}$$

Partial differentiating Equation (6.18) w.r.t. x, it gives

$$\frac{\partial^2 M}{\partial x^2} = \frac{\partial S}{\partial x}.$$

Substituting this equation into Equation (6.14), one obtains

$$(\rho A)\frac{\partial^2 w}{\partial t^2} = \frac{\partial^2 M}{\partial x^2} + p(x). \tag{6.19}$$

From Euler's beam theory, one has

$$M = -\,EI\frac{\partial^2 w}{\partial x^2} \tag{6.20}$$

where I is the second moment of cross-section area of the uniform beam about an axis parallel to the cross-section and through the centroid, and EI is the *flexural rigidity* or *bending stiffness*. The derivation of Equation (6.20) is presented in Appendix 6A.

Substituting Equation (6.20) into (6.19), one obtains

$$(\rho A)\frac{\partial^2 w}{\partial t^2} + (EI)\frac{\partial^4 w}{\partial x^4} = p(x). \tag{6.21}$$

For harmonic motion such that one can assume $w(x,t) = W(x)\sin\omega t$ and free vibration, one can disregard the distributed load so that Equation (6.21) becomes

$$\frac{d^4 W}{dx^4} - \left(\frac{\rho A}{EI}\right)\omega^2\, W = 0.$$

Writing $\beta^4 = \left(\frac{\rho A}{EI}\right)\omega^2$ such that the equation becomes

$$\frac{d^4 W}{dx^4} - \beta^4\, W = 0. \tag{6.22}$$

Note that in Equation (6.22) the differential operator has replaced the partial differential operator because W or $W(x)$ is a function of x only. For conciseness, one writes $D = d/dx$ so that Equation (6.22) can be written as

$$(D^4 - \beta^4)W = 0 \qquad \text{or} \qquad (D^2 + \beta^2)(D^2 - \beta^2)W = 0$$

which gives $D = \pm i\beta,\quad D = \pm\beta$. Therefore, the solution to Equation (6.22) becomes

$$W(x) = B_1 \sin\beta x + B_2 \cos\beta x + B_3 \sinh\beta x + B_4 \cosh\beta x \qquad (6.23)$$

where the arbitrary constants B_1 through B_4 are determined by the boundary conditions. Note that the natural frequencies of lateral or flexural or transverse vibration of a beam is defined, by applying Equation (6.22), as

$$\omega_n = \beta_n^2\sqrt{\frac{EI}{\rho A}}, \qquad n = 1, 2, \ldots, \infty. \qquad (6.24)$$

6.2.1.1 *With standard boundary conditions*

The four arbitrary constants B_1 through B_4 are determined from the end conditions. The *standard* or *simple end conditions* are: (a) simply-supported or pinned or hinged, (b) clamped or fixed, and (c) free. In terms of displacement function $W(x)$ these end conditions are:

(a) *Simply-supported left end conditions*
The deflection and moment are zero. Thus,

$$W(0) = 0, \text{ and } \left.\frac{\mathrm{d}^2W(x)}{\mathrm{d}x^2}\right|_{x=0} = 0. \qquad (6.25a)$$

(b) *Clamped or fixed or built-in left end conditions*
The deflection and rotation are zero. Thus,

$$W(0) = 0, \text{ and } \left.\frac{\mathrm{d}W(x)}{\mathrm{d}x}\right|_{x=0} = 0. \qquad (6.25b)$$

(c) *Free left end conditions*
Moment ad shear force are zero. That is,

$$\left.\frac{\mathrm{d}^2W(x)}{\mathrm{d}x^2}\right|_{x=0} = 0, \quad \text{and} \quad \left.\frac{\mathrm{d}^3W(x)}{\mathrm{d}x^3}\right|_{x=0} = 0. \qquad (6.25c)$$

The boundary conditions in the first of Equation (6.25a) and those in Equation (6.25b) are results of geometry and therefore, they are called *geometric* or *essential* or *imposed boundary conditions.* On the other hand, the boundary conditions in the second of Equation (6.25*a*) and those in Equation (6.25*c*) are results of force and moment balance. Therefore, they are referred to as *natural* or *dynamical* or *additional* boundary conditions.

Example 6.2

Consider a uniform beam of cross-section area A and length L simply-supported at both ends. Find the frequency equation and natural frequencies of this beam.

Solution:

With reference to Equation (6.25*a*), the boundary conditions for this beam at both ends are

$$W(0) = 0, \text{ and } \left.\frac{\mathrm{d}^2 W(x)}{\mathrm{d}x^2}\right|_{x=0} = 0,$$

$$W(L) = 0, \text{ and } \left.\frac{\mathrm{d}^2 W(x)}{\mathrm{d}x^2}\right|_{x=L} = 0.$$

Substituting the conditions at $x = 0$ into Equation (6.23), one has

$$B_2 + B_4 = 0$$

and

$$-\beta^2 B_2 + \beta^2 B_4 = 0.$$

Hence, $B_2 = B_4 = 0$.

Substituting the conditions at $x = L$ into Equation (6.23), one has

$$B_1 \sin \beta L + B_3 \sinh \beta L = 0$$

and

$$-\beta^2 B_1 \sin \beta L + \beta^2 B_3 \sinh \beta L = 0.$$

Writing these last two equations in matrix form, one obtains

$$\begin{bmatrix} \sin \beta L & \sinh \beta L \\ -\beta^2 \sin \beta L & \beta^2 \sinh \beta L \end{bmatrix} \begin{pmatrix} B_1 \\ B_3 \end{pmatrix} = \begin{pmatrix} 0 \\ 0 \end{pmatrix}.$$

The frequency equation is obtained by setting the determinant of the coefficient matrix to zero. Thus,

$$\beta^2 \sin \beta L \sinh \beta L = 0.$$

In order for the lhs to satisfy the rhs of this equation, one requires the frequency equation to be

$$\sin \beta L = 0$$

since for nontrivial β, $\sinh \beta L$ is not zero.

The frequency equation leads to

$$\beta_n L = n\pi, \qquad n = 1, 2, 3, \ldots$$

or applying Equation (6.24), the natural frequencies are given by

$$\omega_n = \beta_n^2 \sqrt{\frac{EI}{\rho A}} = \left(\frac{n\pi}{L}\right)^2 \sqrt{\frac{EI}{\rho A}}, \qquad n = 1, 2, 3, \ldots$$

Example 6.3

Consider a cantilever beam of uniform cross-section area A and length L with the origin at the fixed end. Find the frequency equation of this beam structure.

Solution:

With reference to Equation (6.25*b*), the boundary conditions for the clamped end are

$$W(0) = 0, \text{ and } \left.\frac{\mathrm{d}W(x)}{\mathrm{d}x}\right|_{x=0} = 0$$

while the free-end conditions, with reference to Equation (6.25*c*) are

$$\left.\frac{\mathrm{d}^2W(x)}{\mathrm{d}x^2}\right|_{x=L} = 0, \quad \text{and} \quad \left.\frac{\mathrm{d}^3W(x)}{\mathrm{d}x^3}\right|_{x=L} = 0.$$

Substituting the conditions at $x = 0$ into Equation (6.23), one has

$$B_2 + B_4 = 0 \qquad \text{and} \qquad \beta B_1 + \beta B_3 = 0$$

or

$$B_1 + B_3 = 0.$$

The factor β can be removed from the lhs is due to the fact that it is not equal to zero as the structure is constrained and therefore there is no rigid-body mode.

Substituting the conditions at $x = L$ into Equation (6.23), one obtains

$$-\beta^2 B_1 \sin\beta L - \beta^2 B_2 \cos\beta L + \beta^2 B_3 \sinh\beta L + \beta^2 B_4 \cosh\beta L = 0$$

or

$$-B_1 \sin\beta L - B_2 \cos\beta L + B_3 \sinh\beta L + B_4 \cosh\beta L = 0,$$

and

$$-\beta^3 B_1 \cos\beta L + \beta^3 B_2 \sin\beta L + \beta^3 B_3 \cosh\beta L + \beta^3 B_4 \sinh\beta L = 0,$$

or simply

$$-B_1 \cos\beta L + B_2 \sin\beta L + B_3 \cosh\beta L + B_4 \sinh\beta L = 0.$$

Writing all the four equations in matrix form, one obtains

$$\begin{bmatrix} 0 & 1 & 0 & 1 \\ 1 & 0 & 1 & 0 \\ -\sin\beta L & -\cos\beta L & \sinh\beta L & \cosh\beta L \\ -\cos\beta L & \sin\beta L & \cosh\beta L & \sinh\beta L \end{bmatrix} \begin{pmatrix} B_1 \\ B_2 \\ B_3 \\ B_4 \end{pmatrix} = \begin{pmatrix} 0 \\ 0 \\ 0 \\ 0 \end{pmatrix}.$$

The frequency equation is obtained by setting the determinant of the coefficient matrix of the above equation to zero. Thus, after some algebraic manipulation one has the frequency equation given by

$$\cos\beta L \cosh\beta L + 1 = 0.$$

Example 6.4

Consider a free-free or unconstrained beam of uniform cross-section area A and length L with the origin at the left-hand end. Find the frequency equation of this beam structure.

Solution:

With reference to Equation (6.25*c*), the boundary conditions are

$$\left.\frac{\mathrm{d}^2W(x)}{\mathrm{d}x^2}\right|_{x=0}=0,\quad \text{and} \quad \left.\frac{\mathrm{d}^3W(x)}{\mathrm{d}x^3}\right|_{x=0}=0.$$

Substituting these conditions at $x=0$ into Equation (6.23), one has

$$-\beta^2B_2+\beta^2B_4=0, \qquad -\beta^3B_1+\beta^3B_3=0.$$

Similarly, substituting the conditions at $x=L$ into Equation (6.23), one obtains

$$-\beta^2B_1\sin\beta L-\beta^2B_2\cos\beta L+\beta^2B_3\sinh\beta L+\beta^2B_4\cosh\beta L=0$$

and

$$-\beta^3B_1\cos\beta L+\beta^3B_2\sin\beta L+\beta^3B_3\cosh\beta L+\beta^3B_4\sinh\beta L=0.$$

Note that in the foregoing the factors β^2 and β^3 cannot be removed from the lhs of the equations because β can be equal to zero, associating with the rigid-body motion for the unconstrained structure. Writing all the four equations in matrix form, one obtains

$$\begin{bmatrix} 0 & -\beta^2 & 0 & \beta^2 \\ -\beta^3 & 0 & \beta^3 & 0 \\ -\beta^2\sin\beta L & -\beta^2\cos\beta L & \beta^2\sinh\beta L & \beta^2\cosh\beta L \\ -\beta^3\cos\beta L & \beta^3\sin\beta L & \beta^3\cosh\beta L & \beta^3\sinh\beta L \end{bmatrix}\begin{pmatrix} B_1 \\ B_2 \\ B_3 \\ B_4 \end{pmatrix}=\begin{pmatrix} 0 \\ 0 \\ 0 \\ 0 \end{pmatrix}.$$

The frequency equation is obtained by setting the determinant of the coefficient matrix of the above equation to zero. Thus, after some algebraic manipulation one obtains

$$\beta^2(\cos\beta L\cosh\beta L-1)=0$$

which gives two zero values of β and the remaining values of β are non-zero. These two zero values correspond to the rigid body modes of translation and rotation.

Remarks

It is interesting to note that the frequency equation for a clamped-clamped beam structure applying the same procedure is $\cos\beta L\cosh\beta L-1=$

0. That is, numerically the non-zero natural frequencies of the free-free beam are the natural frequencies of a clamped-clamped beam.

6.2.1.2 *With general boundary conditions*

In addition to the standard or simple boundary conditions discussed in the last sub-section more complex boundary conditions are frequently encountered in practice. For example, lumped mass and springs having translational and rotational restraint at the ends of the beam structures are included. These end conditions often render the orthogonality property of normal modes invalid and therefore specific procedures are designed to formulate the problem at hand to satisfy the orthogonality property. The mathematically rigorous approach seems to be that applied the symbolic or Dirac delta function together with the Lagrangian (Balachandran and Magrab, 2004, 484).

Consider the beam structure with boundaries at $x = 0$ and L. The transverse displacement of the beam is $w = w(x, t)$. The Lagrangian is given by

$$\Gamma = \int_0^L [F_B + G_0\delta_D(x) + G_L\delta_D(x - L)]dx, \tag{6.26}$$

where δ_D denotes the Dirac delta function The first part in the integral (6.26) represents the contributions to the kinetic and potential energy in the beam interior $0 < x < L$, as well as the work done by the external conservative loadings. The function F_B is written in general as

$$F_B = F_B\left(x, t, w, \frac{\partial w}{\partial x}, \frac{\partial w}{\partial t}, \frac{\partial^2 w}{\partial x \partial t}, \frac{\partial^2 w}{\partial x^2}\right),$$

the second and third parts in the integral (6.26) also represent the contributions to the kinetic and potential energy by the discrete elements at the boundaries $x = 0$ and L, respectively, where G_1 and G_2 are functions of $x, t, w, \frac{\partial w}{\partial x}, \frac{\partial w}{\partial t}, \frac{\partial^2 w}{\partial x \partial t}, \frac{\partial^2 w}{\partial x^2}$ in general.

For discussion of the approach, the beam system shown in Figure 6.6 is considered as an example. There is a transverse distributed loading *fc* (*x*, *t*) acting on the uniform beam, and the function F_B is given by

$$F_B = \frac{1}{2}\left[\rho A\left(\frac{\partial\omega}{\partial t}\right)^2 - EJ\left(\frac{\partial^2\omega}{\partial x^2}\right)^2 - k_f\omega^2\right] + f_c\omega,$$

at the left boundary $x = 0$, there is a linear translation spring with stiffness k_1 and a linear torsion spring with stiffness k_{t1}, as well as an inertia element with concentrated mass $M1$ and rotatory inertia $J1$. It is similar for the right boundary $x = L$. The functions G_1 and G_2 are given by

$$G_i = \frac{1}{2}\left[M_i\left(\frac{\partial\omega}{\partial t}\right)^2 + J_i\left(\frac{\partial^2\omega}{\partial x\partial t}\right)^2 - k_i\omega^2 - k_{t_i}\left(\frac{\partial\omega}{\partial x}\right)^2\right], \quad (i = 1,\ 2),$$

where the subscript 1 (or 2) denotes the values taken at the left end $x = 0$ (or the right end $x = L$) of the beam.

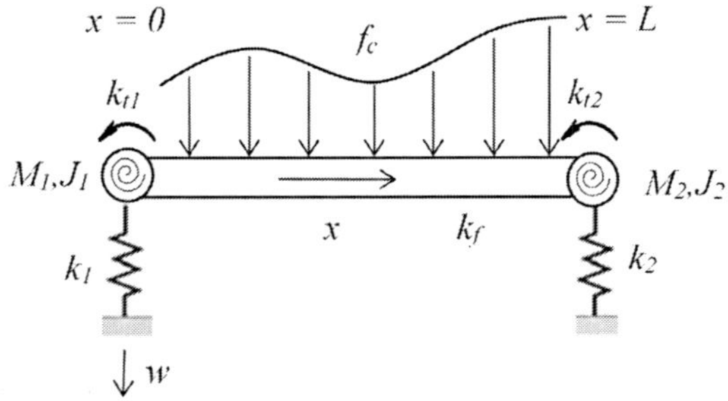

Figure 6.6: Beam structure with general end conditions.

Based on the extended Hamilton's variation principle for conservative systems, the equation of motion and boundary conditions of the beam system in Figure 6.6 are obtained by taking the stationary value of the integral

$$I_H(x,t) = \int_{t_1}^{t_2} \Gamma dt$$

$$= \int_{t_1}^{t_2}\int_0^L [F_B + G_1\delta_D(x) + G_2\delta_D(x-L)]dxdt, \quad (6.27)$$

That is, $\delta I_H(x,t) = 0,$ in which δ is the variation operator. However, in order to obtain the equation of motion and boundary conditions integration by parts has to be performed twice, meaning the amount of algebraic manipulation is comparable to the other approach adopted by Warburton (Warburton, 1976, 117-123, 177-183). The latter approach is applied in this sub-section. The main difference between the approaches in Balachandran and Magrab (Balachandran and Magrab, 2004, 484-486) and Warburton is that the former requires background knowledge of *calculus of variation* whereas the latter only requires a second mathematical course in an undergraduate engineering degree program.

Another example for more general boundary conditions of a uniform beam is shown in Figure 6.7, in which there is an element with concentrated mass m_L, a linear spring of stiffness k_L, and a rotational spring stiffness k_L^R opposing rotation at the starting end, $x = 0$. The boundary conditions at $x = 0$ are given by

$$- EI \left(\frac{\partial^2 w}{\partial x^2}\right)_{x=0} = - k_L^R \left(\frac{\partial w}{\partial x}\right)_{x=0}, \tag{6.28a}$$

and

$$- EI \left(\frac{\partial^3 w}{\partial x^3}\right)_{x=0} = k_L\, w(0,t) + m_L \left(\frac{\partial^2 w}{\partial t^2}\right)_{x=0}, \tag{6.28b}$$

where the subscript L denotes the left-hand end. Note that Equation (6.28*b*) clearly is time-dependent and this is the source of difficulty renders the orthogonality properties being not satisfied if these end conditions are not dealt with properly.

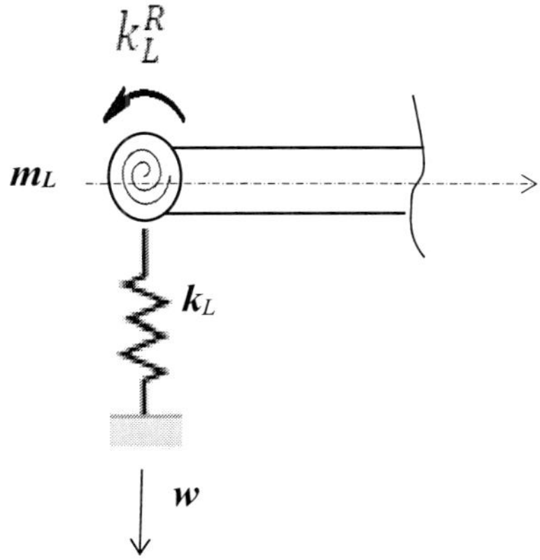

Figure 6.7: Beam structure with concentrated mass and springs.

In order to satisfy the orthogonality property for beam, the normal modes have to be treated as in Appendix 6B. Note that in the case with general boundary conditions, the amount of algebraic manipulation is substantial since integration by parts has to be performed twice. Thus, the amount of algebraic manipulation for the approach by Balachandran and Magrab (Balachandran and Magrab, 2004, 498-502) is essentially similar to that of Warburton (Warburton, 1976, 123,179,181,182), with respect to the amount of algebraic manipulation. Moreover, it is mentioned that if the system is non-conservative the contribution of non-conservative factors (such as damping) should be considered additionally.

Example 6.5

Consider a cantilever beam of a uniform cross-section area A and length L, in which a concentrated mass m_R with its rotary inertia $J_R = m_R\kappa^2$, is attached to the free-end, as shown in Figure 6.8. Given that the boundary conditions are:

$$\text{at} \quad x = 0, \qquad w(x,t) = Be^{i\omega t}, \; \frac{\partial w}{\partial x} = 0, \quad \text{and}$$

$$\text{at} \quad x = L, \quad EI\left(\frac{\partial^2 w}{\partial x^2}\right) = -J_R\left(\frac{\partial^3 w}{\partial t^2 \partial x}\right), \; EI\left(\frac{\partial^3 w}{\partial x^3}\right) = m_R\left(\frac{\partial^2 w}{\partial t^2}\right).$$

Find the frequency equation of this cantilever beam structure.

Solution:

When the harmonic excitation is applied at the base of the cantilever structure, as shown in Figure 6.8, the solution to Equation (6.21) with $p(x) = 0$ is

$$w_o(x,t) = Be^{i\omega t} + w(x,t) \qquad \text{in which} \qquad w(x,t) = W(x)e^{i\omega t}$$

where $W(x) = W(x,\omega)$ is defined in Equation (6.23), and the solution $w_o(x,t)$ is that including the rigid-body displacement $Be^{i\omega t}$. The latter is required to be added to the elastic solution $w(x,t)$ for system with *time-dependent boundary conditions* (Warburton, 1976, 177). However, in the present case and similar cantilever systems in this chapter, only the solution $w(x,t) = W(x)e^{i\omega t}$ is considered.

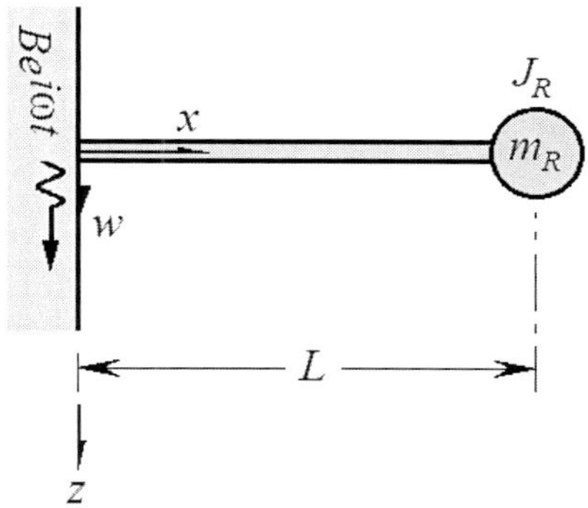

Figure 6.8: Cantilever beam with a concentrated mass and rotary inertia.

Substituting the given boundary conditions into Equation (6.23) successively and going through some algebraic manipulation, one can find

$$W(x,\omega) = H(x,\omega)B \qquad (i)$$

in which B is the amplitude of the applied harmonic force and $H(x,\omega)$ is called the frequency response function or receptance and is given by

$$H(x,\omega) = \frac{1}{2\Delta}\,[N_{1u}\cos\beta x + N_{2u}\cosh\beta x + N_{3u}(\sin\beta x - \sinh\beta x)], \qquad (ii)$$

where

$$\begin{aligned}
N_{1u} &= 1 + \cos u\cosh u - \sin u\sinh u - 2R\cos u\sinh u \\
&\quad + \phi u\,[R\,(1 - \cos u\cosh u + \sin u\sinh u) - 2\sin u\cosh u],\\
N_{2u} &= 1 + \cos u\cosh u + \sin u\sinh u - 2R\sin u\cosh u \\
&\quad + \phi u\,[R\,(1 - \cos u\cosh u - \sin u\sinh u) + 2\cos u\sinh u],\\
N_{3u} &= \sin u\cosh u + \cos u\sinh u + 2\phi u\cos u\cosh u \\
&\quad - 2R\sin u\sinh u - R\phi u\,(\sin u\cosh u + \cos u\sinh u\,),
\end{aligned}$$

and

$$\begin{aligned}
\Delta &= 1 + \cos u\cosh u - \phi u\,(\sin u\cosh u - \cos u\sinh u) \\
&\quad - R(\sin u\cosh u + \cos u\sinh u) - R\phi u(\cos u\cosh u - 1),
\end{aligned}$$

$$R = \phi\left(\frac{\kappa}{L}\right)^2 u^3, \qquad \phi = \frac{m_R}{\rho AL}, \qquad u = \beta L\,.$$

The frequency equation is $\Delta = 0$. When $R = \phi\left(\frac{\kappa}{L}\right)^2 u^3 = 0$ the frequency equation becomes

$$1 + \cos u \cosh u - \phi u\,(\sin u \cosh u - \cos u \sinh u) = 0. \qquad (iii)$$

The above equation is for the case of a cantilever with a concentrated mass attached at its free-end. The only differences are the symbols employed.

Example 6.6
Consider the cantilever beam structure studied in **Example 6.5** to include the tip mass whose centre of gravity is not at the point of attachment. In this case, as shown in Figure 6.9, the boundaries are given as:

$$\text{at} \quad x = 0, \qquad w(x,t) = Be^{i\omega t}, \ \frac{\partial w}{\partial x} = 0, \quad \text{and}$$

$$\text{at} \quad x = L, \quad EI\left(\frac{\partial^2 w}{\partial x^2}\right) = -\,(J_R + m_R\gamma^2)\left(\frac{\partial^3 w}{\partial t^2 \partial x}\right) - m_R\gamma^2\left(\frac{\partial^2 w}{\partial t^2}\right),$$

$$EI\left(\frac{\partial^3 w}{\partial x^3}\right) = m_R\left(\frac{\partial^2 w}{\partial t^2}\right) + m_R\gamma^2\left(\frac{\partial^3 w}{\partial t^2 \partial x}\right), \quad J_R = m_R\kappa^2,$$

where κ is the radius of gyration of the concentrated mass, and γ is the distance between the point of attachment and centre of gravity of the concentrated mass. Find the frequency equation of this cantilever beam structure.

Solution:
When the harmonic excitation is applied at the base of the cantilever structure as shown in Figure 6.9, similar to **Example 6.5**, the solution is

$$w(x,t) = W(x)e^{i\omega t}$$

where $W(x) = W(x,\omega)$ is defined in Equation (6.23).

Substituting the given boundary conditions into Equation (6.23) successively and going through some algebraic manipulation, one finally finds

$$W(x,\omega) = H_1(x,\omega)B \qquad (i)$$

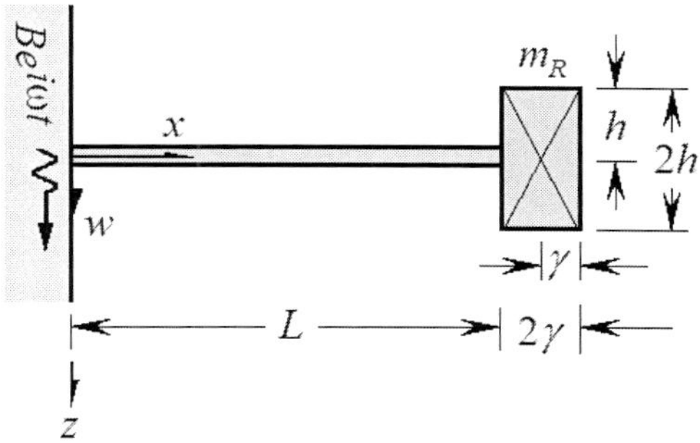

Figure 6.9: Cantilever beam with an extended concentrated mass.

in which $H_1(x,\omega)$ or simply H_1 is given by

$$H_1 = \frac{1}{2\Delta_1}\left[M_{1u}\cos\beta x + M_{2u}\cosh\beta x + M_{3u}(\sin\beta x - \sinh\beta x)\right], \qquad (ii)$$

where

$$\begin{aligned} M_{1u} = {} & 1 + \cos u \cosh u - \sin u \sinh u \\ & - 2\phi u \sin u \cosh u - 2R_1 \cos u \sinh u \\ & + (R_1\phi u - R_2^2)\,(1 - \cos u \cosh u + \sin u \sinh u), \end{aligned}$$

$$\begin{aligned} M_{2u} = {} & 1 + \cos u \cosh u + \sin u \sinh u \\ & + 2\phi u \cos u \sinh u - 2R_1 \sin u \cosh u \\ & + (R_1\phi u - R_2^2)\,(1 - \cos u \cosh u - \sin u \sinh u), \end{aligned}$$

$$\begin{aligned} M_{3u} = {} & (\sin u \cosh u + \cos u \sinh u)(\,1 - R_1\phi u + R_2^2) \\ & -2[\,\phi u \cos u \cosh u - R_1 \sin u \sinh - R_2\,(\sin u \cosh u - \cos u \sinh u)], \end{aligned}$$

$$\begin{aligned} \Delta_1 = {} & 1 + \cos u \cosh u - \phi u\,(\sin u \cosh u - \cos u \sinh u) \\ & - R_1[\sin u \cosh u + \cos u \sinh u + \phi u(\cos u \cosh u - 1)] \\ & - 2R_2 \sin u \sinh u - R_2^2\,(1 - \cos u \cosh u), \end{aligned}$$

and

$$R_1 = \frac{(J_R + m_R\gamma^2)\omega^2}{EI\beta} = \phi\, u^3\left(\frac{\kappa^2+\gamma^2}{L^2}\right), \qquad R_2 = \frac{(m_R\gamma)\omega^2}{EI\beta^2} = \phi u^2\left(\frac{\gamma}{L}\right),$$

$$\phi = \frac{m_R}{\rho AL}, \qquad u = \beta L\,.$$

The frequency equation is

$$\Delta_1 = 0.$$

When the point of attachment of the concentrated mass coincides with the center of gravity of the mass, that is, $\gamma = 0$, Δ_1 is identical to Δ in **Example 6.5**. Table 6.1 includes some representative roots of the frequency equation.

Table 6.1: Representative roots of frequency equation $\Delta_1 = 0$.
Source: To C. W. S. 1982. 83(4), 453.

j	γ/L					
	0.4	0.4	0.4	0.6	0.6	0.8
	κ/L					
	0.6	0.8	1.0	0.8	1.0	1.0
1	0.93735	0.89457	0.85068	0.84425	0.81048	0.77280
2	2.24496	2.08261	1.98013	2.15522	2.04543	2.10370
3	5.02740	4.97253	4.94508	5.02002	4.97823	5.01576
4	8.03827	8.00717	7.99216	8.04124	8.01500	8.04301
5	11.12765	11.10656	11.09652	11.13224	11.11347	11.13504

6.2.2 Lateral vibration of beam with axial forces

The case of lateral vibration of beam with axial forces, as shown in Figure 6.10, can be similarly treated as the sum of the lateral vibration of beam and lateral vibration of cable or string dealt with in Sub-section 6.1.1, since linear vibration is considered.

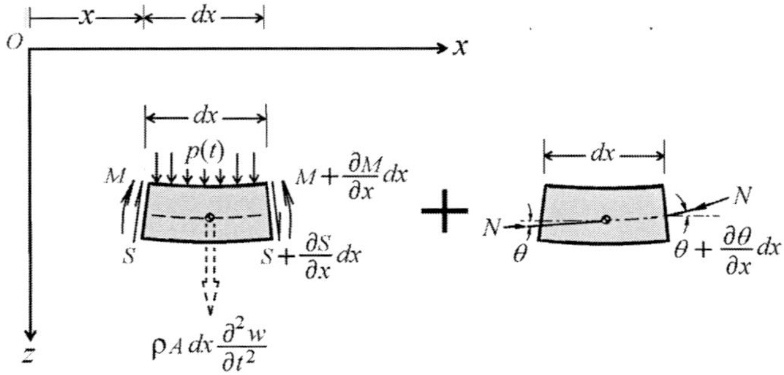

Figure 6.10: Elementary length of beam in bending and with axial forces.

With reference to the FBD in Figure 6.10 and similar to the derivation of Equation (6.17), summing all forces vertically,

$$(\rho A dx)\frac{\partial^2 w}{\partial t^2} = \left(S + \frac{\partial S}{\partial x}dx\right) - S + N\left(\theta + \frac{\partial \theta}{\partial x}dx\right) - N\theta + p(x)dx$$

where N is the axial force or compressive force, $p(x)$ the distributed load or load intensity which is expressed in force per unit length, w or $w(x,t)$ is the transverse or lateral deflection of the beam, and the remaining symbols have their usual meaning. Simplifying the last equation, one has

$$\rho A\frac{\partial^2 w}{\partial t^2} = \frac{\partial S}{\partial x} + N\frac{\partial \theta}{\partial x} + p(x) \tag{6.29}$$

Summing all moments about an axis through the center of mass of the elementary length, and following similar steps in the case of lateral vibration of beam in Sub-section 6.2.1, one arrives at

$$\rho A\frac{\partial^2 w}{\partial t^2} = \frac{\partial^2 M}{\partial x^2} + N\frac{\partial \theta}{\partial x} + p(x). \tag{6.30}$$

Applying the bending moment M in Equation (6.20) and replacing $\theta = \frac{\partial w}{\partial x}$, Equation (6.30) becomes

$$\rho A\frac{\partial^2 w}{\partial t^2} + EI\frac{\partial^4 w}{\partial x^4} - N\frac{\partial^2 w}{\partial x^2} = p(x). \tag{6.31}$$

This is the equation of motion for lateral vibration of beam with axial compressive forces. To provide an illustration of the application of Equation (6.31), one considers the following example.

Example 6.7

Consider a uniform beam of cross-section *A*, length *L* with simply-supported ends, and compressive forces, as shown in Figure 6.11. Find the natural frequencies of this beam structure.

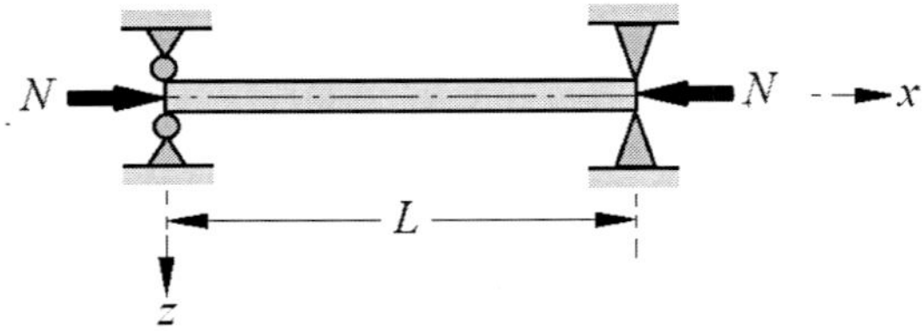

Figure 6.11: Simply-supported uniform beam under compressive axial forces.

Solution:
For harmonic motion, and since damping is not included, a solution which satisfies the given end conditions is

$$w(x,t) = B_1\, sin\left(\frac{n\pi x}{L}\right)\, \sin \omega_n t. \qquad (i)$$

This satisfies Equation (6.31) with $p(x) = 0,$ if

$$-(\rho A)\omega_n^2 + (EI)\left(\frac{n\pi}{L}\right)^4 - N\left(\frac{n\pi}{L}\right)^2 = 0. \qquad (ii)$$

Therefore, the natural frequencies are given by

$$\omega_n = \left(\frac{n\pi}{L}\right)^2 \sqrt{\frac{EI}{\rho A}\left(1 - \frac{NL^2}{n^2\pi^2 EI}\right)}, \qquad n = 1, 2, 3, \ldots. \qquad (iii)$$

Applying this equation, the fundamental mode becomes

$$\omega_1 = \left(\frac{\pi}{L}\right)^2 \sqrt{\frac{EI}{\rho A}\left(1 - \frac{N}{N_c}\right)}, \qquad (iv)$$

where $N_c = \frac{\pi^2 EI}{L^2}$ is known as the Euler critical load for the strut. Buckling of the strut occurs when $N = N_c$ which corresponds to $\omega_1 = 0$.

Before leaving this section, it may be pointed out that when axial tensile forces are applied at the ends of the simply-supported beam in Figure 6.11 instead of compressive forces the natural frequencies are increased since now

$$\omega_n = \left(\frac{n\pi}{L}\right)^2 \sqrt{\frac{EI}{\rho A}\left(1 + \frac{NL^2}{n^2\pi^2 EI}\right)}, \qquad n = 1, 2, 3, ... \qquad (v)$$

6.2.3 Lateral vibration of pipe conveying flowing fluid

Consider a displaced position of the uniform pipe as shown in Figure 6.12. Note that the axial force or tension in the pipe has been disregarded. The fluid has a constant velocity *U* relative to the pipe and tangent to the centre line of the pipe.

For small displacements, the *x*-component and *z*-component of its velocity are, respectively equal to

$$v_x = U\cos\theta = U, \qquad v_z = U\sin\theta = U\theta = U\frac{\partial w}{\partial x}.$$

The point on the center line of the pipe coinciding with the fluid particle has a component of velocity in the *z*-component, $\frac{\partial w}{\partial t}$. Thus, the net velocity of the fluid in the *z*-component becomes

$$V_z = \frac{\partial w}{\partial t} + U\frac{\partial w}{\partial x}.$$

The forces of the fluid in the elementary length of the pipe along the *z*-direction is

$$\mu_f dx\frac{dV_z}{dt} = \mu_f dx\left(U^2\frac{\partial^2 w}{\partial x^2} + 2U\frac{\partial^2 w}{\partial x\partial t} + \frac{\partial^2 w}{\partial t^2}\right),$$

where μ_f is the mass of fluid per unit length contained in the pipe.

The total derivative $\frac{dV_z}{dt}$ is $\frac{dV_z}{dt} = \frac{\partial V_z}{\partial x}\left(\frac{\partial x}{\partial t}\right) + \frac{\partial V_z}{\partial t} = \frac{\partial\left(\frac{\partial w}{\partial t} + U\frac{\partial w}{\partial x}\right)}{\partial x}\left(\frac{\partial x}{\partial t}\right) + \frac{\partial\left(\frac{\partial w}{\partial t} + U\frac{\partial w}{\partial x}\right)}{\partial t} = U^2\frac{\partial^2 w}{\partial x^2} + 2U\frac{\partial^2 w}{\partial x\partial t} + \frac{\partial^2 w}{\partial t^2}$ in which $\frac{\partial x}{\partial t} = U$ has been applied.

Similar to the derivation of Equation (6.17), one sums all forces vertically for the elementary length,

$$(\mu_p dx)\frac{\partial^2 w}{\partial t^2} + \mu_f dx\left(U^2\frac{\partial^2 w}{\partial x^2} + 2U\frac{\partial^2 w}{\partial x\partial t} + \frac{\partial^2 w}{\partial t^2}\right) = \left(S + \frac{\partial S}{\partial x}dx\right) - S.$$

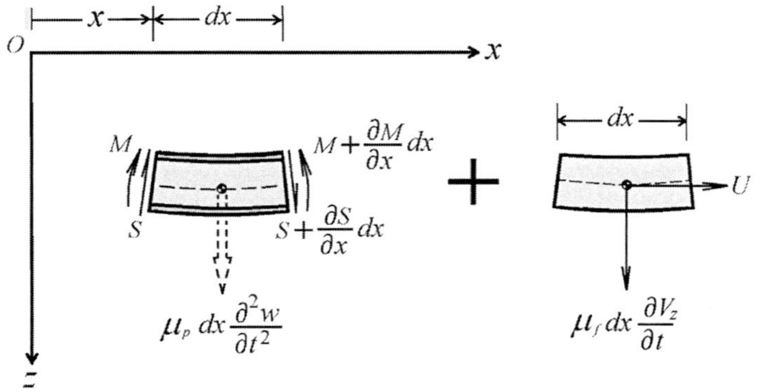

Figure 6.12: Element of the bending pipe with flowing fluid.

After simplification, one has

$$(\mu_p)\frac{\partial^2 w}{\partial t^2}+\mu_f\left(U^2\frac{\partial^2 w}{\partial x^2}+2U\frac{\partial^2 w}{\partial x\partial t}+\frac{\partial^2 w}{\partial t^2}\right)=\frac{\partial S}{\partial x}.$$

Similar to the derivation of Equation (6.19), the rhs term is equal to $-\,EI\left(\frac{\partial^4 w}{\partial x^4}\right)$ and therefore, the last equation becomes

$$(\mu_p)\frac{\partial^2 w}{\partial t^2}+EI\left(\frac{\partial^4 w}{\partial x^4}\right)+\mu_f\left(U^2\frac{\partial^2 w}{\partial x^2}+2U\frac{\partial^2 w}{\partial x\partial t}+\frac{\partial^2 w}{\partial t^2}\right)=0,$$

where μ_p is the mass of the pipe (without the fluid) per unit length of the pipe, and the remaining symbols have already been defined in the foregoing sub-section. The above equation can be re-arranged to give

$$EI\left(\frac{\partial^4 w}{\partial x^4}\right)+\mu_f\left(U^2\frac{\partial^2 w}{\partial x^2}+2U\frac{\partial^2 w}{\partial x\partial t}\right)+(\mu_p+\mu_f)\frac{\partial^2 w}{\partial t^2}=0. \qquad (6.32)$$

This is the equation of motion for a uniform pipe conveying a moving fluid. The first term on the lhs is the elastic force due to bending of the pipe, the term $\mu_f\, U^2\frac{\partial^2 w}{\partial x^2}$ is the centrifugal force of the fluid as it flows in the curved part of the pipe (note that $\frac{\partial^2 w}{\partial x^2}$ is the reciprocal to the local radius of curvature), $2\mu_f U \frac{\partial^2 w}{\partial x \partial t}$ is the Coriolis force since $\frac{\partial^2 w}{\partial x \partial t} = \frac{\partial\left(\frac{\partial w}{\partial x}\right)}{\partial t} = \frac{\partial \theta}{\partial t}$ is the local angular velocity, and the last term is the inertial force of the fluid-containing pipe in the transversal direction.

The solution to Equation (6.32) is complicated by the Coriolis forcing term. In particular, the natural frequencies and mode shapes of this system are generally determined by some numerical approaches. Typical examples can be found in the excellent book by Païdoussis (Païdoussis, 2014, 531-537). However, exact solution to the eigenvalue problem was provided by Kaladi (Kaladi, 1984, 116-124).

A solution to Equation (6.32) can be written as

$$w(x,t) = W(x,\omega)e^{i\omega t} \tag{6.33}$$

where the spatially dependent term,

$$W(x,\omega) = B_1 e^{\gamma_1 x} + B_2 e^{\gamma_2 x} + B_3 e^{\gamma_3 x} + B_4 e^{\gamma_4 x} \tag{6.34}$$

in which B_1 through B_4 are arbitrary constants to be found by applying the boundary conditions, and γ_1 through γ_4 are the roots of the characteristic equation

$$(EI)\gamma^4 + (\mu_f U^2)\gamma^2 + i(2U\omega\mu_f)\gamma - (\mu_p + \mu_f)\omega^2 = 0, \tag{6.35}$$

which may be obtained by first assuming $w(x,t) = Ce^{\gamma x}e^{i\omega t}$ where C is a constant, and then applying Equation (6.32). This is a quartic polynomial in γ and the exact solution can be obtained by a standard procedure. Owing to the third term on the lhs of Equation (6.35), the values for γ are complex in general.

Differentiating Equation (6.34) with respect to *x*,

$$\frac{dW(x,\omega)}{dx} = W' = B_1\gamma_1 e^{\gamma_1 x} + B_2\gamma_2 e^{\gamma_2 x} + B_3\gamma_3 e^{\gamma_3 x} + B_4\gamma_4 e^{\gamma_4 x}, \tag{6.36a}$$

$$\frac{d^2W(x,\omega)}{dx^2} = W'' = B_1\gamma_1^2 e^{\gamma_1 x} + B_2\gamma_2^2 e^{\gamma_2 x} + B_3\gamma_3^2 e^{\gamma_3 x} + B_4\gamma_4^2 e^{\gamma_4 x}, \tag{6.36b}$$

$$\frac{d^3W(x,\omega)}{dx^3} = W''' = B_1\gamma_1^3 e^{\gamma_1 x} + B_2\gamma_2^3 e^{\gamma_2 x} + B_3\gamma_3^3 e^{\gamma_3 x} + B_4\gamma_4^3 e^{\gamma_4 x}. \tag{6.36c}$$

Writing the spatial depend rotation, moment, and shear forces, respectively as

$$\Theta(x,\omega) = \Theta = -W' = -\frac{dW(x,\omega)}{dx},$$

$$\mathbb{M}(x,\omega) = \mathbb{M} = -EIW'',$$

$$\mathbb{S}(x,\omega) = \mathbb{S} = -EIW''',$$

and applying Equation (6.34), one can have the following state vector equation at x,

$$Z(x) = G(x)\mathbb{B} \tag{6.37}$$

in which

$$Z(x) = \begin{pmatrix} -W \\ \Theta \\ \mathbb{M} \\ \mathbb{S} \end{pmatrix}, \quad \mathbb{B} = -\begin{pmatrix} B_1 \\ B_2 \\ B_3 \\ B_4 \end{pmatrix}, \quad G(x) = \begin{bmatrix} g_{11} & g_{12} & g_{13} & g_{14} \\ g_{21} & g_{22} & g_{23} & g_{24} \\ g_{31} & g_{32} & g_{33} & g_{34} \\ g_{41} & g_{42} & g_{43} & g_{44} \end{bmatrix},$$

Where

$$g_{11} = e^{\gamma_1 x}, \quad g_{12} = e^{\gamma_2 x}, \quad g_{13} = e^{\gamma_3 x}, \quad g_{14} = e^{\gamma_4 x},$$

$$g_{21} = \gamma_1 e^{\gamma_1 x}, \quad g_{22} = \gamma_2 e^{\gamma_2 x}, \quad g_{23} = \gamma_3 e^{\gamma_3 x}, \quad g_{24} = \gamma_4 e^{\gamma_4 x},$$

$$g_{31} = EI\gamma_1^2 e^{\gamma_1 x}, \quad g_{32} = EI\gamma_2^2 e^{\gamma_2 x}, \quad g_{33} = EI\gamma_3^2 e^{\gamma_3 x}, \quad g_{34} = EI\gamma_4^2 e^{\gamma_4 x},$$

$$g_{41} = EI\gamma_1^3 e^{\gamma_1 x}, \quad g_{42} = EI\gamma_2^3 e^{\gamma_2 x}, \quad g_{43} = EI\gamma_3^3 e^{\gamma_3 x}, \quad g_{44} = EI\gamma_4^3 e^{\gamma_4 x}.$$

At $x = 0$, Equation (6.37) gives

$$Z(0) = G(0)\mathbb{B}, \qquad \mathbb{B} = [G(0)]^{-1}Z(0).$$

By making use of the last equation, therefore, the state vector equation for $x = L$ becomes

$$Z(L) = G(L)[G(0)]^{-1}Z(0) = \mathbb{T}\,Z(0), \tag{6.38}$$

where the transfer matrix, $\mathbb{T} = G(L)[G(0)]^{-1}$ relates the state vectors at two ends of the pipe containing the moving fluid.

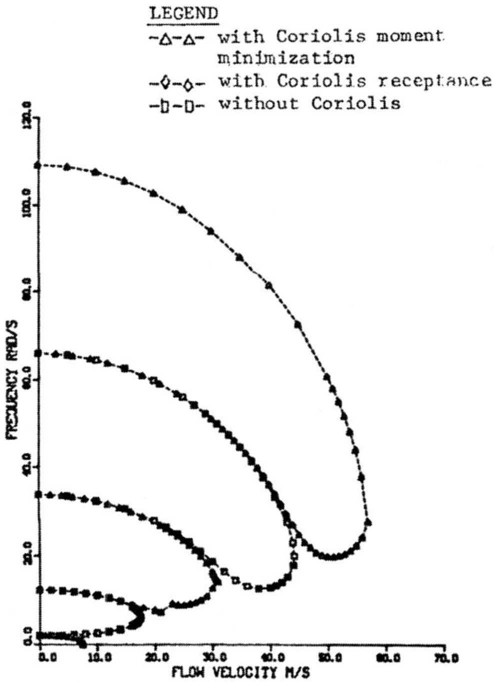

Figure 6.13: Frequency plots. *Source:* Kaladi, 1984, 61

Since γ_1 through γ_4 are functions of ω and therefore, the transfer matrix $\mathbb{T}$ is also a function of ω. It can be evaluated at any given value of ω, as long as the properties of the pipe and fluid are known. This can be achieved by a computer program. Then, by making use of Equation (6.38), natural frequencies of a particular pipe conveying fluid with given end conditions can be determined. Once the natural frequencies are determined Equation (6.34) can be applied to evaluate the mode shapes of vibration. Again, a computer program has been written to perform such calculation. Computed results for a cantilever pipe with fluid are presented in Figures 6.13 and 6.14. The geometrical and material properties of this cantilever pipe are included in Table 6.2. More computed results were provided in (Kaladi, 1984, 59-67).

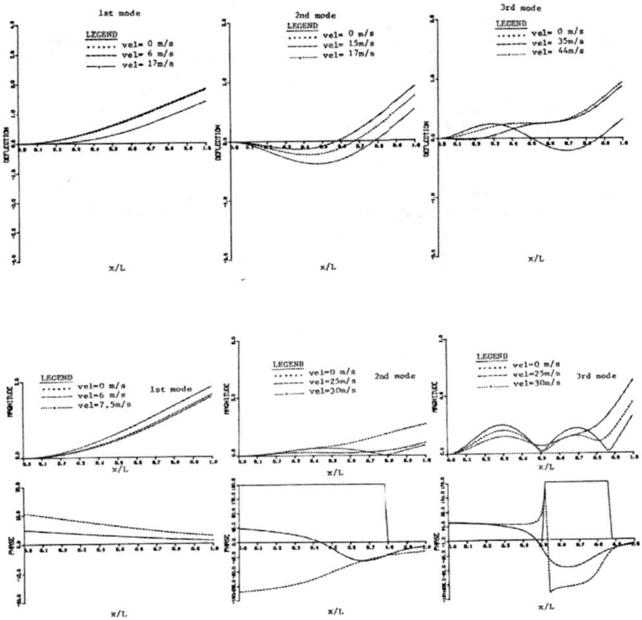

Figure 6.14: Mode shapes. *Source:* Kaladi, 1984, 66, 67

Table 6.2: Geometrical and material properties of cantilever pipe
Source: Kaladi, 1984, 45

Young's Modulus	5.519 x 10^9 N/m^2
Pipe inside diameter	17.7 mm
Pipe wall thickness	1.9588 mm
Pipe length	2.52984 m
Mass of pipe per unit length	1.6668 kg/m
Mass of fluid per unit length of pipe	0.23811 kg/m

6.2.4 Lateral vibration of rotating beam

The lateral vibration of a rotating beam is important in the vibration analysis of helicopter blades, for example. In this case, the theoretical treatment is very similar to that presented Sub-section 6.2.2 for the lateral vibration of a beam with an axial force. For simplicity, in the following the outline of the derivation of the equation of motion of a rotating beam of uniform cross-section is presented.

Consider the elementary length of the beam shown in Figure 6.15 in which the *inertia force* or *centrifugal force* (which is opposite in direction to the *centripetal acceleration*) is given by

$$F_c = \int_x^L (\rho A)\Omega^2 \, \xi d\xi = \frac{1}{2} (\rho A)\Omega^2 \, L^2 \left(1 - \frac{x^2}{L^2}\right), \tag{6.39}$$

where Ω is the rotating speed of the beam, and the remaining symbols have already defined in the foregoing sub-sections. Note, it has been assumed that the radius of the vertical shaft in Figure 6.15 is small relative to the length of the rotating beam BC.

With reference to the FBD in Figure 6.15 and similar to the derivation of Equation (6.17), summing all forces vertically,

$$(\rho A dx)\frac{\partial^2 w}{\partial t^2} = \left(S + \frac{\partial S}{\partial x} dx\right) - S + F_c\left(\theta + \frac{\partial \theta}{\partial x} dx\right) - F_c\theta + p(x)dx$$

where F_c replaces the axial force or tension in the derivation of Equation (6.29).

Simplifying the last equation, one has

$$(\rho A)\frac{\partial^2 w}{\partial t^2} = \frac{\partial S}{\partial x} + F_c \frac{\partial \theta}{\partial x} + p(x) \tag{6.40}$$

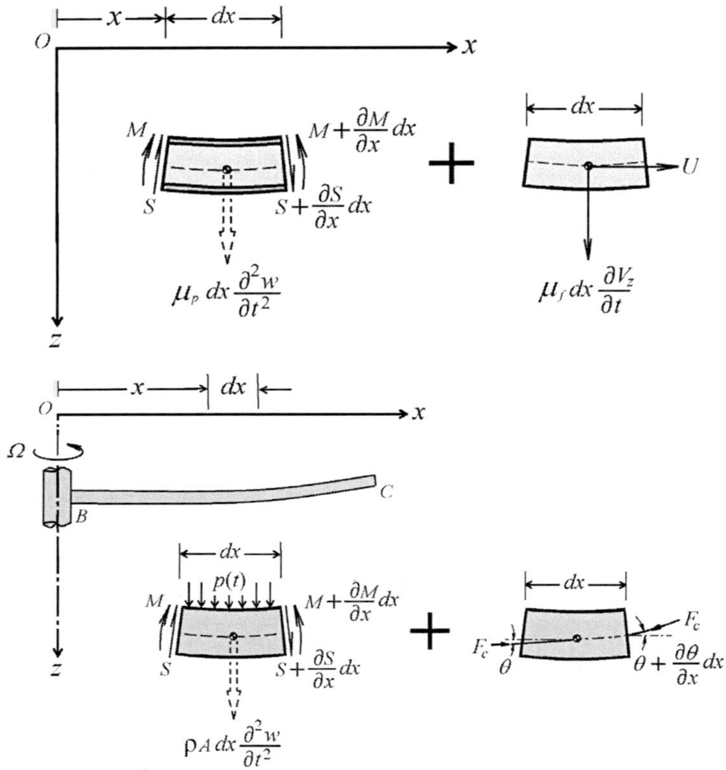

Figure 6.15: Elementary length of rotating uniform beam.

Summing all moments about an axis through the center of mass of the elementary length, and following similar steps in the case of the lateral vibration of a beam in Sub-section 6.2.1, one arrives at

$$(\rho A)\frac{\partial^2 w}{\partial t^2} = \frac{\partial^2 M}{\partial x^2} + F_c \frac{\partial \theta}{\partial x} + p(x). \qquad (6.41)$$

Applying the bending moment M in Equation (6.20) and replacing $\theta = \frac{\partial w}{\partial x}$, Equation (6.41) becomes

$$(\rho A)\frac{\partial^2 w}{\partial t^2} + (EI)\frac{\partial^4 w}{\partial x^4} - \frac{1}{2}(\rho A)\Omega^2 L^2\left(1 - \frac{x^2}{L^2}\right)\frac{\partial^2 w}{\partial x^2} = p(x). \qquad (6.42)$$

This is the equation of motion for the lateral vibration of a rotating beam and the boundary conditions are similar to those for a stationary beam in lateral vibration.

To provide an illustration of the application of Equation (6.42), one considers the following example.

Example 6.8

Consider the helicopter blade as a uniform cantilever beam structure rotating at Ω. The boundary is given as:

at $x = 0,\quad w(x,t) = B\,sin\omega t\,,\quad \frac{\partial w}{\partial x} = 0,\quad$ and

at $x = L,\quad EI\left(\frac{\partial^2 w}{\partial x^2}\right) = 0,\quad$ and $\quad EI\left(\frac{\partial^3 w}{\partial x^3}\right) = 0.$

Find the frequency equation of this rotating cantilever beam structure.

Solution:

When the harmonic excitation is applied at the base or hub of the helicopter blade and since the structure has no damping, one can assume a solution

$$w(x,t) = W(x,\omega)sin\omega t = C\,e^{rx}sin\omega t \qquad (i)$$

where $W(x,\omega) = C\,e^{rx}$ and C is an arbitrary constant.

Applying this general solution to Equation (6.43) and disregarding the rhs, one obtains the auxiliary equation

$$-\left(\frac{\rho A}{EI}\right)\omega^2 + r^4 - \frac{1}{2}\left(\frac{\rho A}{EI}\right)\Omega^2 L^2\left(1 - \frac{x^2}{L^2}\right)r^2 = 0.$$

Simplifying,

$$r^4 - \left(\frac{N}{EI}\right)r^2 - \left(\frac{\rho A}{EI}\right)\omega^2 = 0, \qquad (ii)$$

where $N = N(x) = \frac{1}{2}(\rho A)\Omega^2 L^2\left(1 - \frac{x^2}{L^2}\right)$.

Solving Equation (*ii*),

$$r^2 = \frac{N}{2EI} \pm \sqrt{\left(\frac{N}{2EI}\right)^2 + \left(\frac{\rho A}{EI}\right)\omega^2},$$

and therefore, the roots of the auxiliary equation are

$$r_1 = \pm\sqrt{\sqrt{\left(\frac{N}{2EI}\right)^2 + \left(\frac{\rho A}{EI}\right)\omega^2} + \frac{N}{2EI}},$$

$$r_2^* = \pm\, i\, r_2 = \pm\, i\sqrt{\sqrt{\left(\frac{N}{2EI}\right)^2 + \left(\frac{\rho A}{EI}\right)\omega^2} - \frac{N}{2EI}}.$$

Substituting these roots into Equation (*i*), one obtains

$$W(x,\omega) = C_1 sin(r_1 x) + C_2 cos(r_1 x) + C_3 sinh(r_2 x) + C_4 cosh(r_2 x), \qquad (iii)$$

where the arbitrary constants are determined by the given boundary conditions.

Before proceeding further, one first considers the centrifugal force

$$N = N(x) = \frac{1}{2}\,(\rho A)\Omega^2 L^2 \left(1 - \frac{x^2}{L^2}\right).$$

At $x = 0,\ N_0 = N(0) = \frac{1}{2}\,(\rho A)\Omega^2 L^2$, and at $x = L,\ N_1 = N(L) = 0$

With these two values,

$$r_{10} = r_1|_{x=0} = \pm\sqrt{\sqrt{\left(\frac{N_0}{2EI}\right)^2 + \left(\frac{\rho A}{EI}\right)\omega^2} + \frac{N_0}{2EI}},$$

$$r_{11} = r_1|_{x=L} = \pm\sqrt{\sqrt{\left(\frac{\rho A}{EI}\right)\omega^2}} = \pm\beta,$$

$$r_{20} = \pm\sqrt{\sqrt{\left(\frac{N_0}{2EI}\right)^2 + \left(\frac{\rho A}{EI}\right)\omega^2} - \frac{N_0}{2EI}},$$

$$r_{21} = \pm = \pm\sqrt{\sqrt{\left(\frac{\rho A}{EI}\right)\omega^2}} = \pm\beta = r_{11},$$

in which β has been defined by Equation (6.19).

Now, applying the given boundary conditions and after some lengthy algebraic manipulation, one obtains the frequency equation as

$$M_{1r}M_{4r} + M_{2r}M_{3r} = 0 \qquad (iv)$$

Where

$$M_{1r} = \beta^2\left[\left(\frac{r_{20}}{r_{10}}\right)\sin u + \sinh u\right], \qquad M_{2r} = \beta^2(\cos u + \cosh u),$$
$$M_{3r} = \beta^3\left[\left(\frac{r_{20}}{r_{10}}\right)\cos u + \cosh u\right], \qquad M_{4r} = \beta^3(\sin u - \sinh u),$$

$$u = \beta L.$$

Operating on Equation (iv) and simplifying, the frequency equation becomes

$$\left(1 + \frac{r_{20}}{r_{10}}\right)(1 + \cos u \cosh u) + \left(1 - \frac{r_{20}}{r_{10}}\right)\sin u \sinh u = 0. \qquad (v)$$

When the helicopter blade is stationary, meaning $\Omega = 0$ and $\frac{r_{20}}{r_{10}} = 1$, Equation (iv) reduces to

$$1 + \cos u \cosh u = 0 \qquad (vi)$$

which is the frequency equation for a stationary cantilever beam.

6.2.5 Lateral vibration of beam with elastic foundation

Similar to the derivation of equation of motion for a beam in lateral vibration presented in Sub-section 6.2.1, one considers an elementary length of the beam on an elastic foundation, as shown in Figure 6.16. The elastic foundation is represented by the distributed linear springs whose modulus is $k_f = E_f I_f$ which is the load on unit length of the foundation required to produce unit deflection.

Summing all forces vertically, one has

$$(\rho A dx)\frac{\partial^2 w}{\partial t^2} = \left(S + \frac{\partial S}{\partial x}dx\right) - S - (k_f)wdx + p(x)dx$$

where the symbols have been defined in Sub-section 6.2.1. Simplifying, one has

$$(\rho A)\frac{\partial^2 w}{\partial t^2} = \frac{\partial S}{\partial x} - (k_f)w + p(x).$$

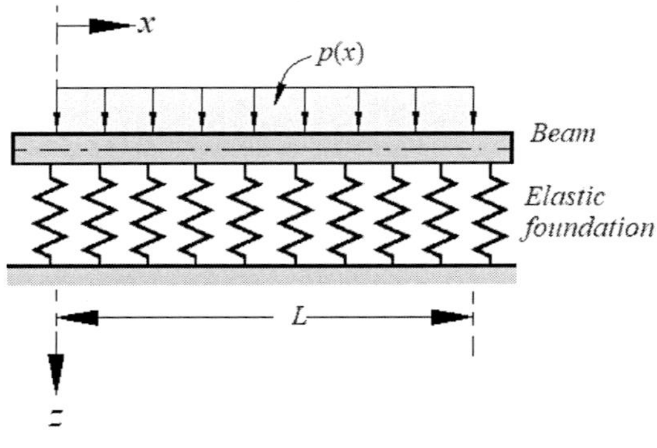

Figure 6.16: Beam on elastic foundation.

Then, following the steps for the derivation of Equation (6.21) one can show that the equation of motion becomes

$$(\rho A)\frac{\partial^2 w}{\partial t^2} + (EI)\frac{\partial^4 w}{\partial x^4} + (k_f)w = p(x). \tag{6.43}$$

For harmonic motion such that one can assume $w(x,t) = W(x)\sin\omega t$ and free vibration, one can disregard the distributed load so that Equation (6.21) becomes

$$\frac{\partial^4 W}{\partial x^4} - \left(\frac{\rho A \omega^2 - k_f}{EI}\right) W = 0.$$

Writing $\beta_1^4 = \left(\frac{\rho A \omega^2 - k_f}{EI}\right)$ such that the equation becomes

$$\frac{d^4 W}{dx^4} - \beta_1^4 W = 0. \tag{6.44}$$

Similar to the derivation of Equation (6.23), the solution to Equation (6.44) is

$$W(x) = B_1 \sin\beta_1 x + B_2 \cos\beta_1 x + B_3 \sinh\beta_1 x + B_4 \cosh\beta_1 x \tag{6.45}$$

where the arbitrary constants B_1 through B_4 are determined by the boundary conditions. The natural frequencies of lateral or flexural or transverse vibration of a beam on an elastic foundation is defined, by applying Equation (6.44), as

$$\omega_n = \beta_{1n}^2 \sqrt{\frac{EI + \frac{k_f}{\beta_{1n}^4}}{\rho A}}, \qquad n = 1, 2, 3, \ldots, \infty. \tag{6.46}$$

The contribution of the elastic foundation is to increase the natural frequency of the corresponding Euler beam in lateral vibration.

For free vibration analysis, a beam on elastic foundation can be considered as a free-free structure since, in general, the stiffness of the foundation is much lower than that of the beam. Thus, the frequency equation can be similarly obtained as that for the free-free beam in **Example 6.4**. That is,

$$\beta_1^2(\cos\beta_1 L \cosh\beta_1 L - 1) = 0$$

which gives two zero values of β_1 and the remaining values of β_1 are non-zero. The two zero values correspond to the rigid body modes of translation and rotation for the beam structure.

6.2.6 Lateral vibration of Timoshenko beam

The beam considered in this sub-section includes for both the effects of shear deformation and rotary inertia of the beam. The total deflection $w(x,t)$ of the beam comprises of two components, one due to bending and the other due to shear. Thus, the slope of the deflection curve at x, $\frac{\partial w}{\partial x}$ is defined as the sum of the angle θ of rotation due to bending and the angle α_s of distortion due to shear, as shown in Figure 6.17(*a*) in which the uniformly distributed load has been disregarded for clarity of illustration. In other words,

$$\frac{\partial w}{\partial x} = \theta + \alpha_s. \tag{6.47a}$$

(*a*)

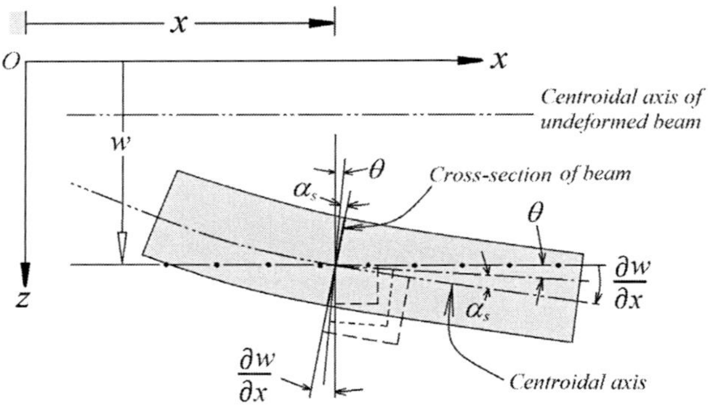

Figure 6.17: (*a*) Kinematics in Timoshenko beam.

(b)

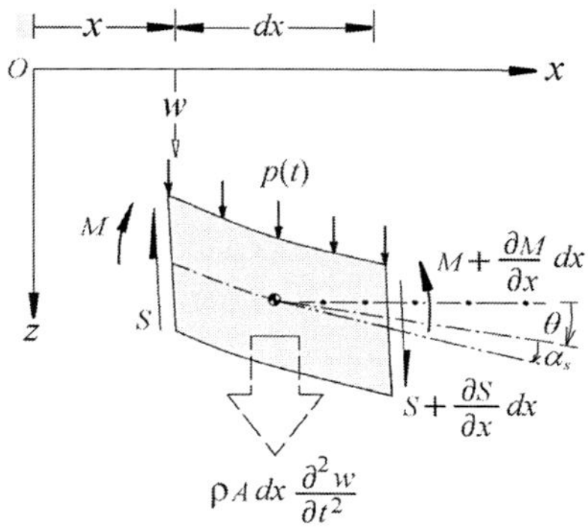

Figure 6.17: (*b*) elementary length of Timoshenko beam.

An elementary length of the Timoshenko beam is illustrated in Figure 6.17(*b*) in which the slope of the deflection curve is at $x+\frac{dx}{2}$, strictly speaking. However, $\frac{dx}{2}$ approaches zero, in the limit, and therefore, the slopes $\frac{\partial w}{\partial x}$ at $x+\frac{dx}{2}$ and at x are considered equal.

The shear force is given by Timoshenko and associate (Timoshenko and Gere, 1961, 132-134) as

$$S=\kappa AG\alpha_s \quad \text{such that} \quad \frac{\partial w}{\partial x}-\theta=\frac{S}{\kappa AG}, \tag{6.47b}$$

where κ is usually defined as the ratio of the average shear strain over the cross-section to the shear strain at the centroid and it depends on the shape of the cross-section as well as strictly on the mode of vibration, A is the cross-sectional area, and G the shear modulus. The values of κ for several cross-sections by Cowper (Cowper, 1966, 33(2), 335-340) are adopted in the present discussion. Equations (6.47a) (6.47b) are similar to those in Warburton (Warburton, 1976, 204), Meirovitch (Meirovitch, 1967, 129), and Timoshenko and Gere (Timoshenko and Gere, 1961, 132-134).

The moment-curvature relation is

$$M=-EI\frac{\partial\theta}{\partial x}. \tag{6.48}$$

Similar to transverse vibration of Euler beam in Section 6.2.1, one sums all the forces vertically,

$$(\rho A dx)\frac{\partial^2 w}{\partial t^2} = \left(S + \frac{\partial S}{\partial x}dx\right) - S + p(x)dx$$

where the symbols have already been defined in the foregoing. Simplifying the last equation, one has

$$(\rho A)\frac{\partial^2 w}{\partial t^2} = \frac{\partial S}{\partial x} + p(x). \tag{6.49}$$

This is identical in form to Equation (6.17).

Now, summing all moments about an axis through the center of mass of the elementary length as for the Euler beam in Section 6.2.1, and including rotary inertia, one obtains

$$(\rho I dx)\frac{\partial^2 \theta}{\partial t^2} = M + S\left(\frac{dx}{2}\right) + \left(S + \frac{\partial S}{\partial x}dx\right)\left(\frac{dx}{2}\right) - \left[M(x) + \frac{\partial M}{\partial x}dx\right]$$

in which ρI is the mass moment of inertia per unit length about an axis through the centroid of the elementary length of the beam. After simplifying,

$$(\rho I)\frac{\partial^2 \theta}{\partial t^2} = S - \frac{\partial M}{\partial x}. \tag{6.50}$$

Substituting Equations (6.47) and (6.48) into (6.50),

$$(\rho I)\frac{\partial^2 \theta}{\partial t^2} + \frac{\partial\left(-EI\frac{\partial \theta}{\partial x}\right)}{\partial x} - \kappa AG\left(\frac{\partial w}{\partial x} - \theta\right) = 0.$$

Rearranging terms,

$$(\rho I)\frac{\partial^2 \theta}{\partial t^2} - EI\frac{\partial^2 \theta}{\partial x^2} + \kappa AG\left(\theta - \frac{\partial w}{\partial x}\right) = 0. \tag{6.51}$$

By making use of Equations (6.47) and (6.49), one can obtain

$$(\rho A)\frac{\partial^2 w}{\partial t^2} - \kappa AG\,\frac{\partial\left(\frac{\partial w}{\partial x} - \theta\right)}{\partial x} = p(x) \qquad \text{or}$$

$$(\rho A)\frac{\partial^2 w}{\partial t^2} - \kappa AG\left(\frac{\partial^2 w}{\partial x^2} - \frac{\partial \theta}{\partial x}\right) = p(x). \tag{6.52}$$

Equations (6.51) and (6.52) are two coupled equations of motion for the beam.

By partially differentiating Equation (6.51) w.r.t. x, one obtains

$$\left[(\rho I)\frac{\partial^2}{\partial t^2} - EI\frac{\partial^2}{\partial x^2} + \kappa AG\right]\left(\frac{\partial \theta}{\partial x}\right) = \kappa AG\left(\frac{\partial^2 w}{\partial x^2}\right). \tag{6.53}$$

From Equation (6.52), one has

$$\frac{\partial \theta}{\partial x} = \frac{p(x)}{\kappa AG} - \left(\frac{\rho A}{\kappa AG}\right)\frac{\partial^2 w}{\partial t^2} + \frac{\partial^2 w}{\partial x^2}\,. \tag{6.54}$$

Substituting this into Equation (6.53), performing the expansion, simplifying, and rearranging, one arrives at

$$EI\frac{\partial^4 w}{\partial x^4} - \left(\rho I + \frac{EI\rho}{\kappa G}\right)\frac{\partial^4 w}{\partial x^2 \partial t^2} + \rho A\frac{\partial^2 w}{\partial t^2} + \left(\frac{I\rho^2}{\kappa G}\right)\frac{\partial^4 w}{\partial t^4}$$
$$= p(x) + \left(\frac{I\rho}{\kappa AG}\right)\frac{\partial^2 p(x)}{\partial t^2} - \left(\frac{EI}{\kappa AG}\right)\frac{\partial^2 p(x)}{\partial x^2}\,. \tag{6.55}$$

Equation (6.55) is the governing equation of motion for a uniform beam including the effects of shear deformation and rotary inertia. It is generally known as the *Timoshenko beam equation*. Clearly, the *Euler beam equation*, Equation (6.21) is a special case of the Timoshenko beam equation.

Equation (6.55) can be further simplified if $p(x) = 0$ which corresponds to the case of free vibration and writing $I = A\gamma^2$ where γ is the radius of gyration,

$$\frac{\partial^4 w}{\partial x^4} - \rho\left(\frac{1}{E} + \frac{1}{\kappa G}\right)\frac{\partial^4 w}{\partial x^2 \partial t^2} + \frac{\rho}{E\gamma^2}\frac{\partial^2 w}{\partial t^2} + \left(\frac{\rho^2}{\kappa GE}\right)\frac{\partial^4 w}{\partial t^4} = 0\,. \tag{6.56}$$

This equation agrees with Equation (4.65) of Warburton (1976, 205), except for the symbols.

For harmonic motion such that one can assume $w(x,t) = Be^{\lambda x}\sin\omega t$ then the following equation will satisfy Equation (6.56),

$$\lambda^4 + \left(\frac{1}{c_L^2} + \frac{1}{\kappa c_S^2}\right)(\lambda\omega)^2 - \frac{1}{\gamma^2 c_L^2}\omega^2 + \frac{1}{\kappa c_L^2 c_S^2}\omega^4 = 0\,, \qquad (6.57)$$

in which

$$c_L^2 = \frac{E}{\rho}, \quad c_S^2 = \frac{G}{\rho}$$

with c_L being analogous to the longitudinal wave propagation of the rod defined in Equation (6.8) and c_s being analogous to the velocity of propagation of shear waves in the rod defined in Equation (6.13). Equation (6.57) agrees with Equation (4.66) of Warburton (1976, 205), except for the symbols used. Equation (6.57) can further be written in a more convenient way as

$$\lambda^4 + b\lambda^2 + c = 0\,, \qquad (6.58)$$

where $b = \left(\frac{1}{c_L^2} + \frac{1}{\kappa c_S^2}\right)\omega^2$, $c = \frac{\omega^4}{\kappa c_L^2 c_S^2} - \frac{\omega^2}{\gamma^2 c_L^2}$. In general when $\frac{\gamma^2\omega^2}{\kappa c_S^2} < 1$ which corresponds to the low frequency case (note that the high frequency case is when $\frac{\gamma^2\omega^2}{\kappa c_S^2} > 1$). Equation (6.58) yields one negative and one positive value of λ^2 which means the roots are $\pm i\lambda_1,\ \pm\lambda_2$. Then one has

$$w(x,t) = W(x)\sin\omega t \qquad (6.59)$$

Where

$$W(x) = B_1 \sin\lambda_1 x + B_2\cos\lambda_1 x + B_3\sinh\lambda_2 x + B_4\cosh\lambda_2 x.$$

The four arbitrary constants B_1 through B_4 are determined from the end conditions. The *standard* or *simple end conditions* for Timoshenko beams are similar to those for the Euler beam with due modifications to rotation of cross-section area and shear deformation. Specifically, they are included in the following:

(a) *Simply-supported left end conditions*
The deflection and moment are zero. Thus,

$$w(0,t) = 0, \quad \text{and} \quad \left.\frac{\partial\theta(x,t)}{\partial x}\right|_{x=0} = 0. \tag{6.60a}$$

(*b*) *Clamped or fixed or built-in left end conditions*
The deflection and rotation are zero. Thus,

$$w(0,t) = 0, \text{ and } \quad \theta(x,t)|_{x=0} = 0. \tag{6.60b}$$

(*c*) *Free left end conditions*
Moment ad shear force are zero. That is,

$$\left.\frac{\partial\theta(x,t)}{\partial x}\right|_{x=0} = 0, \qquad \text{and} \quad S = \kappa AG\left(\frac{\partial w}{\partial x} - \theta\right)\bigg|_{x=0} = 0. \tag{6.60c}$$

The boundary conditions in the first of Equation (6.60*a*) and those in Equation (6.60*b*) are results of geometry and therefore, they are called *geometric* or *essential* or *imposed boundary conditions.* On the other hand, the boundary conditions in the second of Equation (6.60*a*) and those in Equation (6.60*c*) are results of force and moment balance. Therefore, they are referred to as *natural* or *dynamical* or *additional* boundary conditions.

Example 6.9
Consider a uniform Timoshenko beam of cross-section area A and length L simply-supported at both ends. Find the frequency equation of this beam.

Solution:
With reference to Equation (6.60*a*), the boundary conditions for this beam at both ends are

$$w(0,t) = 0, \text{ and } \quad \left.\frac{\partial\theta(x,t)}{\partial x}\right|_{x=0} = 0,$$

$$w(L,t) = 0, \text{ and } \quad \left.\frac{\partial\theta(x,t)}{\partial x}\right|_{x=L} = 0.$$

For the first set of end conditions at $x = 0$, substituting $w(0,t) = 0$ into Equation (6.59), one has

$$B_2 + B_4 = 0. \tag{i}$$

The second condition $\left.\frac{\partial\theta(x,t)}{\partial x}\right|_{x=0} = 0$, from Equation (6.47*a*), one has

$$\theta = \frac{\partial w}{\partial x} - \alpha_s$$

such that by performing partial differentiation of the latter w.r.t. x leads to

$$\frac{\partial \theta}{\partial x} = \frac{\partial^2 w}{\partial x^2} - \frac{\partial \alpha_s}{\partial x}$$

which upon application of Equation (6.47b) becomes

$$\frac{\partial \theta}{\partial x} = \frac{\partial^2 w}{\partial x^2} - \left(\frac{1}{\kappa AG}\right)\frac{\partial S}{\partial x}.$$

By making use of Equations (6.49) and (6.59), the last equation reduces to

$$\frac{\partial \theta}{\partial x} = \frac{\partial^2 w}{\partial x^2} - \left(\frac{\rho\omega^2}{\kappa G}\right) w.$$

Therefore, at $x = 0$ and applying Equation (6.59) to the above equation, it gives

$$-\lambda_1^2 B_2 + \lambda_2^2 B_4 = 0. \qquad (ii)$$

Solving these two equations, one finds $B_2 = B_4 = 0$.

Substituting the conditions at $x = L$ into Equation (6.59), one has

$$B_1 \sin \lambda_1 L + B_3 \sinh \lambda_2 L = 0 \qquad (iii)$$

and

$$-\lambda_1^2 B_1 \sin \lambda_1 L + \lambda_2^2 B_3 \sinh \lambda_2 L = 0. \qquad (iv)$$

Multiplying Equation (iii) by λ_1^2 and adding the resulting equation to (iv), one has $B_3 = 0$. Substituting the latter into Equation (iii), one obtains $\sin \lambda_1 L = 0$ which leads to $\lambda_1 = n\pi/L$ with $n = 1, 2, 3, \ldots$

Substituting $\lambda = i\lambda_1 = in\pi/L$ into Equation (6.58) and after simplifying as well as rearranging, the frequency equation of the simply supported uniform Timoshenko beam becomes

$$\omega^4 - \left[b_1 \left(\frac{n\pi}{L}\right)^2 + b_2\right]\omega^2 + c_o = 0\,, \qquad (v)$$

where $b_1 = c_L^2 + \kappa c_S^2$, $b_2 = \kappa c_S^2/\gamma^2$, and $c_o = \kappa c_L^2 c_S^2 \left(\frac{n\pi}{L}\right)^4$.

It may be appropriate to note that the above frequency equation agrees with the corresponding equation, Equation (4.68) in Warburton (1976, 206). The latter equation was claimed there to be obtained by replacing $\lambda = \lambda_1 = n\pi/L$ in Equation (4.66). However, if such a substitution of $\lambda = \lambda_1$ is followed the positive sign in front of b_1 inside the square brackets on the lhs of Equation (v) will change to negative.

6.3 Normal Modes and Responses of Continuous Systems

In this section the response of a representative continuous system, the uniform Euler beam is considered. While the following presentation is for a particular continuous system the procedure is similarly applicable to other distributed parameter models with due modification to the symbols and their meanings.

Recall, in the derivation of Equation (6.21) the distributed load is a function of space and time. Therefore, $p(x)$ is strictly, $p(x,t)$. Thus, for the current presentation it is written as

$$\rho A \frac{\partial^2 w}{\partial t^2} + EI \frac{\partial^4 w}{\partial x^4} = p(x,t).$$

To include damping of viscoelastic material, one is required to modify the bending moment-curvature relation, Equation (6.20) to

$$M = -EI\left(\frac{\partial^2 w}{\partial x^2} + \delta \frac{\partial^3 w}{\partial x^2 \partial t}\right)$$

where δ is the damping constant which has the dimension of time and not to be confused with the variational operator symbol.

By making use of Equation (6.19), the equation of motion becomes

$$\rho A \frac{\partial^2 w}{\partial t^2} + EI\left(\frac{\partial^4 w}{\partial x^4} + \delta \frac{\partial^4 w}{\partial x^4 \partial t}\right) = p(x,t). \qquad (6.61)$$

The solution to the homogeneous differential equation, that is, with $p(x,t) = 0$, and $\delta = 0$, can be expressed in terms of the normal modes, $W_r(x)$ every of which is associated with a natural frequency ω_r. Thus,

$$w(x,t) = \sum_r B_r W_r(x) \sin(\omega_r t + \varphi_r). \tag{6.62}$$

This is known as the general solution for *free vibration* of the beam.

The general solution to Equation (6.61) may be expressed in the form

$$w(x,t) = \sum_r W_r(x) q_r(t), \tag{6.63}$$

where $q_r(t)$ is a function of time only and is known as the *principal coordinate*.

By applying Equation (6.63) and substituting all the terms into Equation (6.61), multiplying mode shape $W_s(x)$, and integrating with respect to x over the length of the beam,

$$\begin{aligned}\int_0^L (\rho A) W_s(x) \sum_r W_r(x) \frac{d^2 q_r(t)}{dt^2}\, dx \\ + \int_0^L W_s(x)\,(EI) \sum_r \left[\frac{d^4 W_r(x)}{dx^4} q_r(t) \right. \\ \left. + \delta \frac{d^4 W_r(x)}{dx^4} \frac{d q_r(t)}{dt} \right] dx \\ = \int_0^L W_s(x)\, p(x,t) dx. \end{aligned} \tag{6.64}$$

By the orthogonality conditions, Equations (6*B*.11) and (6*B*.13),

$$(\rho A) \int_0^L W_s(x)\, W_r(x)\, dx = 0, \quad (EI) \int_0^L W_s(x) \frac{d^4 W_r(x)}{dx^4}\, dx = 0, \quad r \neq s.$$

In addition, from Equation (6*B*.18),

$$(EI) \int_0^L W_s(x) \frac{d^4 W_s(x)}{dx^4}\, dx = \omega_s^2 (\rho A) \int_0^L W_s^2(x)\, dx$$

so that Equation (6.64) becomes

$$\frac{d^2q_s(t)}{dt^2}+\delta\omega_s^2\frac{dq_s(t)}{dt}+\omega_s^2q_s(t)=\frac{\int_0^L W_s(x)\ p(x,t)dx}{\rho A\int_0^L W_s^2(x)\ dx}. \quad (6.65)$$

Assuming the distributed load, $p(x,t)$ can be written as $p(x,t)=P(x)f(t)$ and the mode shapes are known, then Equation (6.65) reduces to

$$\frac{d^2q_s(t)}{dt^2}+\delta\omega_s^2\frac{dq_s(t)}{dt}+\omega_s^2q_s(t)=\frac{f(t)\int_0^L W_s(x)\,P(x)dx}{\rho A\int_0^L W_s^2(x)\ dx}=Q_s\,f(t), \quad (6.66)$$

in which the ratio, Q_s on the rhs of Equation (6.66) can be found. The latter equation is of the same form as that for a single dof system. Therefore, if the uniform beam is at rest in its equilibrium position at $t=0$ the solution to Equation (6.66) is given by the Duhamel integral as

$$q_s(t)=\frac{Q_s}{\omega_s^d}\int_0^t f(\tau)e^{-\frac{1}{2}\delta\omega_s^2(t-\tau)}\sin\omega_s^d(t-\tau)d\tau \quad (6.67)$$

where $\omega_s^d=\omega_s\sqrt{1-\left(\frac{1}{2}\delta\omega_s\right)^2}$. Once $q_s(t)$ in Equation (6.67) is determined it is substituted into Equation (6.63) to obtain the dynamic response of the beam.

As indicated in Equation (6.63), the dynamic response of the beam depends on the contributions of normal modes. In order to improve on the convergence of the principal coordinates, a method developed by Williams (Warburton, 1976, 66-68) is employed here. This method is known as the *modal acceleration method*. It starts with integrating the rhs of Equation (6.67) by parts so that

$$q_s(t)=\frac{Q_sf(t)}{\omega_s^2}-\frac{Q_s}{\omega_s^2}\int_0^t\frac{df(\tau)}{d\tau}e^{-\frac{1}{2}\delta\omega_s^2(t-\tau)}$$

$$\left[\cos\omega_s^d(t-\tau)+\frac{\delta\omega_s^2}{2\omega_s^d}\sin\omega_s^d(t-\tau)\right]d\tau. \quad (6.68)$$

This is achieved first by considering the integral in Equation (6.6)

$$\int_0^t f(\tau) e^{-\frac{1}{2}\delta\omega_s^2(t-\tau)} \sin\omega_s^d(t-\tau) d\tau.$$

Writing $u = f(\tau)$, and $dv = e^{-\frac{1}{2}\delta\omega_s^2(t-\tau)} \sin\omega_s^d(t-\tau) d\tau$ so that integration by parts gives

$$\int u dv = uv - \int v du$$

where $v = \int dv = \int e^{-\frac{1}{2}\delta\omega_s^2(t-\tau)} \sin\omega_s^d(t-\tau) d\tau$. To operate on the last integral, one can make use of the formula in item 14.518 of Mathematical Handbook of formulas and Tables of Spiegel (Spiegel, 1968, 85). That is,

$$v = \int e^{ax} \sin bx \, dx = \frac{e^{ax}(a \sin bx - b \cos bx)}{a^2 + b^2}$$

in which $a = -\frac{1}{2}\delta\omega_s^2$, $b = \omega_s^d, x = t - \tau$ such that $dx = -d\tau$.

Upon substituting, operating on the integral, and simplifying, one finds

$$\int_0^t e^{-\frac{1}{2}\delta\omega_s^2(t-\tau)} \sin\omega_s^d(t-\tau)\, d\tau$$
$$= \frac{e^{-\frac{1}{2}\delta\omega_s^2(t-\tau)}\omega_s^d}{\omega_s^2}\left[\cos\omega_s^d(t-\tau) + \frac{\delta\omega_s^2}{2\omega_s^d}\sin\omega_s^d(t-\tau)\right].$$

This is applied to the integration of the rhs of Equation (6.67) by parts to obtain the result in Equation (6.68).

6.4 Forced Vibrations of Continuous Systems

In this section, forced vibrations of continuous systems are considered to primarily illustrate the steps in the solution and therefore, only representative cases are presented. In Sub-section 6.4.1, lateral vibration of cable connecting two equal masses, and, in Sub-section 6.4.2, response of cable under initial displacement is considered. Sub-section 6.4.3 is concerned with the forced vibration of a bar under initial displacement while in Sub-section 6.4.4 the steady state response of a taut cable is included. The response of a uniform beam to harmonic force is considered in Sub-section 6.4.5 whereas the response of a simply supported uniform beam to a transient force applied at mid-span of the beam is dealt with in Sub-section 6.4.6. In Sub-section 6.4.7, the response of a uniform beam to a uniformly distributed force is included to illustrate the case in which no normal modes are required. The response of the uniform beam due to a travelling point load is presented in Sub-section 6.4.8. To conclude this section, remarks are included in Sub-section 6.4.9.

6.4.1 Lateral motion of cable connecting two equal masses

A satellite considered as a system that consists of two equal masses m each, connected by a cable of length $2L$. This simplified model is shown in Figure 6.18(a). The system rotates in space with a constant angular velocity Ω. If the variation in the cable tension is disregarded one wish to show that the differential equation of lateral motion of the cable is

$$\frac{\partial^2 y}{\partial x^2} = \frac{\mu}{mL\Omega^2}\left(\frac{\partial^2 y}{\partial t^2} - \Omega^2 y\right),$$

where $y(x,t)$ or y is the displacement or deformation of the cable and μ is the length density (that is, mass per unit length) of the cable.

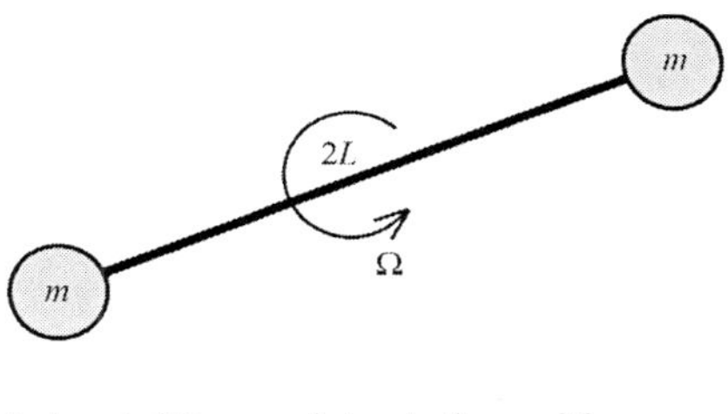

Figure 6.18: (a) A satellite model rotating with an angular velocity.

To derive the equation of lateral motion, one first considers the kinematics of the satellite as shown in Figure 6.18(b). Note that when the angular velocity is zero the system still has vibration and therefore, one still has motion, y and $\frac{\partial^2 y}{\partial t^2}$. Thus, if one assumes the mode shape or motion pattern as that given in Figure 6.18(b) the acceleration in the y-direction is $\frac{\partial^2 y}{\partial t^2} - \Omega^2 y$ for $\Omega \neq 0$. The following is concerned with the derivation of this quantity.

The velocity at point B of the cable is

$$\vec{v}_B = \vec{v}_A + \vec{v}_{B/A}.$$

But the differentiation of the relative velocity with respect to time *t* is

$$\frac{d\vec{v}_{B/A}}{dt} = \frac{d\vec{\Omega}\times\vec{r}}{dt} = \frac{d\vec{\Omega}}{dt}\times\vec{r} + \vec{\Omega}\times\frac{d\,\vec{r}}{dt}.$$

Since $\vec{\Omega}$ is constant and the first term on the rhs becomes zero therefore,

$$\frac{d\vec{v}_{B/A}}{dt} = \vec{\boldsymbol{\Omega}}\times\frac{d\,\vec{r}}{dt}$$

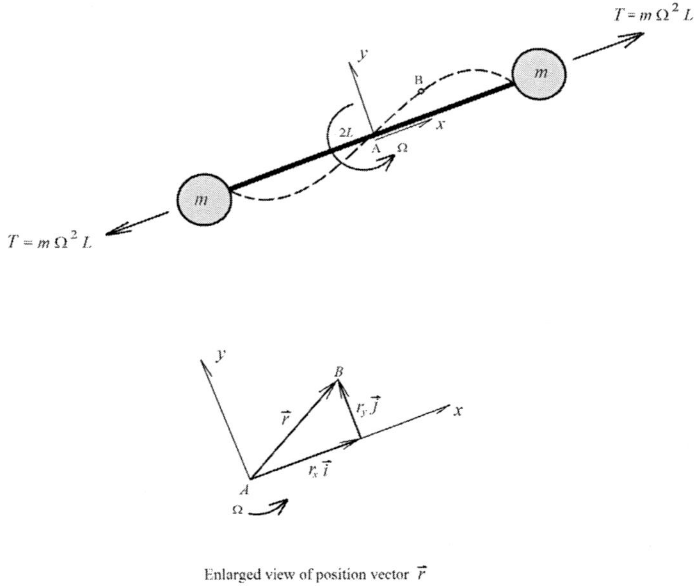

Figure 6.18: (b) Kinematics of satellite model.

where $\frac{d\,\vec{r}}{dt} = r_x\frac{d\vec{\imath}}{dt} + r_y\frac{d\vec{\jmath}}{dt} = \vec{\Omega}\times\left(r_x\,\vec{\imath} + r_y\,\vec{\jmath}\right) = \vec{\Omega}\times\vec{r}$. Note that A*xyz* is the rotating frame of reference in which the *z*-axis is perpendicular to the *x*-

y plane and pointing outward from the x-y plane in accordance with the right-hand screw rule. Thus,

$$\frac{d\vec{v}_{B/A}}{dt} = \vec{\Omega} \times (\vec{\Omega} \times \vec{r}).$$

Operating on this equation, one has

$$\begin{aligned}\frac{d\vec{v}_{B/A}}{dt} &= \vec{\Omega} \times (\vec{\Omega} \times \vec{r}) = \Omega\,\vec{k} \times [\Omega\,\vec{k} \times (r_x\,\vec{\imath} + r_y\,\vec{\jmath})] \\ &= \Omega\,\vec{k} \times (\Omega\,r_x\,\vec{\jmath} - \Omega\,r_y\,\vec{\imath}\,) = \Omega^2(-\,r_x\vec{\imath} - \,r_y\,\vec{\jmath}\,).\end{aligned}$$

Note that according to the symbol given in the example $r_x = x$ and $r_y = y$. Since in the present example only lateral motion is required. That is, only the component along the y-direction is of interest. This component is in addition to the acceleration term $\frac{\partial^2 y}{\partial t^2}$ in which $\Omega = 0$. In other words, the total acceleration in the y-direction is now $\frac{\partial^2 y}{\partial t^2} - \Omega^2 y$.

With reference to Equation (6.1), one has

$$T\frac{\partial^2 y}{\partial x^2} = \mu\left(\frac{\partial^2 y}{\partial t^2} - \Omega^2 y\right).$$

Since the tension in the cable is $T = mL\Omega^2$ therefore this equation becomes

$$\frac{\partial^2 y}{\partial x^2} = \frac{\mu}{mL\Omega^2}\left(\frac{\partial^2 y}{\partial t^2} - \,\Omega^2 y\right).$$

This is the required equation.

6.4.2 Response of cable under initial displacement

A uniform cable of length L and high initial tension is statically displaced h units from the center as shown in Figure 6.19. The cable is released so that it undergoes oscillation. The governing equation of motion in the cable becomes

$$c^2\frac{\partial^2 v(x,t)}{\partial x^2} = \frac{\partial^2 v(x,t)}{\partial t^2}$$

in which $v(x,t)$ or v is the displacement or deformation of the cable (not to be confused with the symbol for velocity in the foregoing sub-section) in the y-direction and the remaining symbols have their usual meaning.

For this problem the boundary conditions are: at $x = 0$, $v(0,t) = 0$ and at $x = L$, $v(L,t) = 0$. The initial conditions are: at $t = 0$,

$$v(x,0) = \begin{cases} \frac{2hx}{L}, & 0 \leq x \leq \frac{L}{2} \\ \frac{2h}{L}(L-x), & \frac{L}{2} \leq x \leq L, \end{cases}$$

$$\frac{\partial v(x,t)}{\partial t}\Big|_{t=0} = 0.$$

Determine the expression for the displacement $v(x,t)$of the cable.

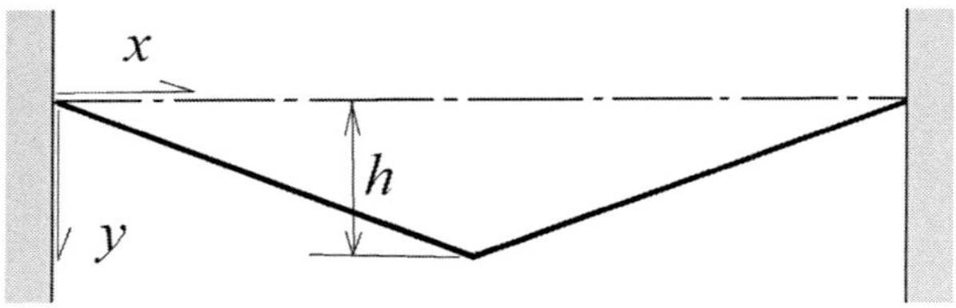

Figure 6.19: Cable being given an initial input displacement.

With reference to Equations (6.1) and (6.2), the deflection of the cable is given by

$$v(x,t) = \left[A \sin\left(\frac{\omega x}{c}\right) + B \cos\left(\frac{\omega x}{c}\right)\right](C \sin \omega t + D cos\omega t), \quad (6.69)$$

where A, B, C, and D are arbitrary constants.

For the present case, the boundary conditions are: at $x = 0$, $v(0,t) = 0$ and at $x = L$, $v(L,t) = 0$. These imply that $B = 0$ and

$$sin\left(\frac{\omega}{c}L\right) = 0 \quad \text{which gives} \quad \frac{\omega_n L}{c} = n\pi, \quad \text{with } n = 1, 2, 3, \ldots.$$

By definition, $c = f\lambda$ where f is the frequency of oscillation and λ is the wavelength. Also, one can write

$$\frac{\omega_n}{c} = \frac{n\pi}{L}$$

so that Equation (6.69) can be written as

$$v(x,t) = \sum_{n=1}^{\infty} \sin\left(\frac{n\pi x}{L}\right)(C_n \sin \omega_n t + D_n \cos \omega_n t). \tag{6.70}$$

The arbitrary constants in this equation are to be determined in the following.

Since the initial conditions are given by

$$v(x,0) = \begin{cases} \dfrac{2hx}{L}, & 0 \le x \le \dfrac{L}{2} \\ \dfrac{2h}{L}(L-x), & \dfrac{L}{2} \le x \le L, \end{cases}$$

$$\left.\frac{\partial v(x,t)}{\partial t}\right|_{t=0} = 0,$$

the latter leads to

$$\left.\frac{\partial v(x,t)}{\partial t}\right|_{t=0} = 0 = \sum_{n=1}^{\infty} C_n \sin\left(\frac{n\pi x}{L}\right).$$

In order that this equation is satisfied $C_n = 0$.

The first initial condition above requires that

$$\sum_{n=1}^{\infty} D_n \sin\left(\frac{n\pi x}{L}\right) = \begin{cases} \dfrac{2hx}{L}, & 0 \le x \le \dfrac{L}{2} \\ \dfrac{2h}{L}(L-x), & \dfrac{L}{2} \le x \le L. \end{cases}$$

In order to determine D_n one multiplies both sides of this equation by $\sin\left(\frac{n\pi x}{L}\right)$ and integrating between the limits $x = 0$ and $x = L$. That is,

$$\int_0^L \sum_{n=1}^{\infty} D_n \sin\left(\frac{n\pi x}{L}\right) \sin\left(\frac{n\pi x}{L}\right) dx = \int_0^{\frac{L}{2}} \left(\frac{2hx}{L}\right) \sin\left(\frac{n\pi x}{L}\right) dx$$
$$+ \int_{\frac{L}{2}}^{L} \left[\frac{2h}{L}(L-x)\right] \sin\left(\frac{n\pi x}{L}\right) dx. \qquad (6.71)$$

Applying the orthogonality property to the lhs of this equation, one obtains

$$\int_0^L \sum_{n=1}^{\infty} D_n \sin\left(\frac{n\pi x}{L}\right) \sin\left(\frac{n\pi x}{L}\right) dx$$
$$= \int_0^L D_n \sin\left(\frac{n\pi x}{L}\right) \sin\left(\frac{n\pi x}{L}\right) dx$$
$$= \int_0^L \frac{D_n}{2}\left[1 - \cos\left(\frac{2n\pi x}{L}\right)\right] dx = \frac{LD_n}{2}. \qquad (6.72)$$

The rhs of Equation (6.71) is

$$\left(\frac{2h}{L}\right) \int_0^{\frac{L}{2}} x \sin\left(\frac{n\pi x}{L}\right) dx + \int_{\frac{L}{2}}^{L} \left(\frac{2h}{L}\right)(L-x) \sin\left(\frac{n\pi x}{L}\right) dx$$
$$= \left(\frac{2h}{L}\right) \int_0^{\frac{L}{2}} x \sin\left(\frac{n\pi x}{L}\right) dx + 2h \int_{\frac{L}{2}}^{L} \sin\left(\frac{n\pi x}{L}\right) dx$$
$$- \left(\frac{2h}{L}\right) \int_{\frac{L}{2}}^{L} x \sin\left(\frac{n\pi x}{L}\right) dx. \qquad (6.73)$$

Now, the second term on the rhs of Equation (6.73) becomes

$$2h \int_{\frac{L}{2}}^{L} \sin\left(\frac{n\pi x}{L}\right) dx = \frac{2hL}{n\pi}\left[\cos\left(\frac{n\pi}{2}\right) - \cos(n\pi)\right]. \qquad (6.74)$$

The first and third terms on the rhs of Equation (6.73) can be evaluated using the technique of integration by parts. Thus, by writing $u = x$, and $dv = \sin\left(\frac{n\pi x}{L}\right) dx$ so that the integral of the first term on the rhs of Equation (6.73) becomes

$$\int_0^{\frac{L}{2}} x \sin\left(\frac{n\pi x}{L}\right) dx = \left[-\frac{Lx}{n\pi}\cos\left(\frac{n\pi x}{L}\right)\right]_0^{L/2} + \frac{L}{n\pi}\int_0^{\frac{L}{2}} \cos\left(\frac{n\pi x}{L}\right) dx$$

$$\int_0^{\frac{L}{2}} x \sin\left(\frac{n\pi x}{L}\right) dx = \left(\frac{L}{n\pi}\right)^2 \sin\left(\frac{n\pi}{2}\right) - \frac{L^2}{2n\pi}\cos\left(\frac{n\pi}{2}\right).$$

Therefore, the first term on the rhs of Equation (6.73) becomes

$$\frac{2h}{L}\int_0^{\frac{L}{2}} x \sin\left(\frac{n\pi x}{L}\right) dx = \frac{2h}{L}\left[\left(\frac{L}{n\pi}\right)^2 sin\left(\frac{n\pi}{2}\right) - \frac{L^2}{2n\pi} cos\left(\frac{n\pi}{2}\right)\right]. \quad (6.75)$$

Similarly, the third term on the rhs of Equation (6.73) can be determined as

$$\frac{2h}{L}\int_{\frac{L}{2}}^{L} x\, sin\left(\frac{n\pi x}{L}\right) dx$$
$$= \frac{2h}{L}\left[\frac{L^2}{2n\pi} cos\left(\frac{n\pi}{2}\right) - \frac{L^2}{2n\pi} cos(n\pi)\right.$$
$$\left. - \left(\frac{L}{n\pi}\right)^2 sin\left(\frac{n\pi}{2}\right)\right]. \quad (6.76)$$

Applying Equations (6.73), (6.74), and (6.76), and simplifying, the rhs of Equation (6.73) becomes

$$\frac{4hL}{(n\pi)^2} sin\left(\frac{n\pi}{2}\right).$$

Equating the above results for Equations (6.73) and (6.75) to Equation (6.71), one has

$$D_n = \frac{8h}{(n\pi)^2} sin\left(\frac{n\pi}{2}\right).$$

Thus,

$$D_n = (-1)^{\frac{n-1}{2}} \frac{8\,h}{n^2\pi^2} \qquad \text{with } n = 1, 3, 5, \ldots$$

This can be rewritten as

$$D_{2r-1} = (-1)^{r-1} \frac{8\,h}{(2r-1)^2\pi^2} \qquad \text{with } r = 1, 2, 3, \ldots$$

Substituting this expression into Eq. (6.70) and with due modification to the integer, one has the required deflection

$$v(x,t) = \sum_{r=1}^{\infty} \frac{(-1)^{r-1}\,8\,h}{(2r-1)^2\pi^2} \sin\left[\frac{(2r-1)\pi x}{L}\right] \cos\left[\frac{(2r-1)\pi\,ct}{L}\right]. \quad (6.77)$$

6.4.3 Response of uniform bar under initial displacement

A simplified model of a rocket may be considered as a uniform bar of length L and its propulsion system may be represented as a force that generates a movement $V_o \cos \omega t$. In practice, the change of mass of the rocket is substantial due to fuel consumption in the process. However, for simplicity, the mass of the rocket is being considered constant within a relatively short duration. This simple model is shown in Figure 6.20.

Assuming a small oscillation, one is interested in finding the steady-state vibration and the applied frequencies ω that cause resonance of the system. To find the steady-state vibration, one reference to the equation for longitudinal vibration of the rocket along the x-direction is given by

$$\frac{\partial^2 v}{\partial t^2} = c^2 \frac{\partial^2 v}{\partial y^2} \quad (6.78)$$

where $v = v(y,t)$ is the displacement of the cross-section of the uniform rocket at time t and position y.

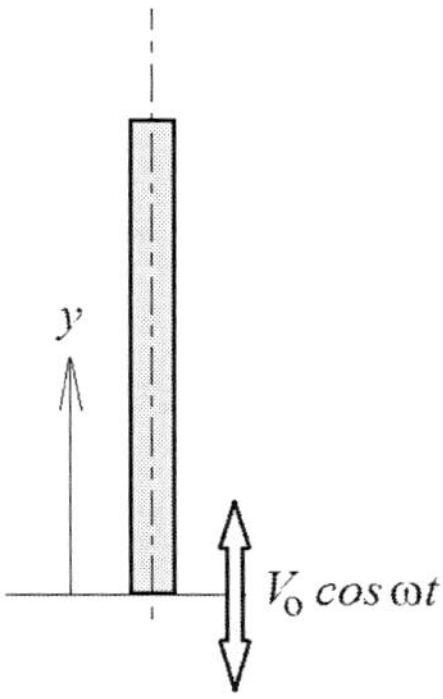

Figure 6.20: A simple model of a rocket.

The boundary conditions are: at $y = 0$, $v(0,t) = V_o \cos \omega t$ and at $y = L$, $\frac{\partial v(y,t)}{\partial y}\Big|_{y=L} = 0$. This latter boundary condition is due to the fact that there is no applied force at the free end of the rocket so that there is no stress and, in turn, no strain.

Since steady state vibration is required and the boundary condition at $y = 0$ is $v(0,t) = V_o \cos \omega t$ therefore one can assume $v(y,t) = V(y) \cos \omega t$ for the small oscillation.

Substituting the small oscillation expression into Equation (6.78), it results in

$$-\omega^2 V(y) \cos \omega t = c^2 \cos \omega t \frac{\partial^2 V(y)}{\partial y^2} \quad \text{or} \quad \frac{d^2 V(y)}{dy^2} + \left(\frac{\omega^2}{c^2}\right) V(y) = 0.$$

The solution to this second order differential equation can be expressed as

$$V(y) = A \sin\left(\frac{\omega y}{c}\right) + B \cos\left(\frac{\omega y}{c}\right).$$

Thus, the general solution for the steady state vibration becomes

$$v(y,t) = \left[A \sin\left(\frac{\omega y}{c}\right) + B \cos\left(\frac{\omega y}{c}\right)\right] \cos \omega t. \tag{6.79}$$

From the first boundary condition,

$$v(0,t) = V_o \cos \omega t = B \cos \omega t \qquad \text{which gives } B = V_o\,.$$

From the second boundary condition,

$$\left.\frac{\partial v(y,t)}{\partial y}\right|_{y=L} = 0 = \frac{\omega}{c}\left[A\cos\left(\frac{\omega L}{c}\right) - V_o\sin\left(\frac{\omega L}{c}\right)\right]\cos\omega t.$$

This leads to $A = V_o\tan\left(\frac{\omega L}{c}\right)$. Thus, the steady state vibration is

$$v(y,t) = V_o\left[\tan\left(\frac{\omega L}{c}\right)\sin\left(\frac{\omega y}{c}\right) + \cos\left(\frac{\omega y}{c}\right)\right]\cos\omega t. \qquad (6.80)$$

This equation indicates that $v(y,t)\rightarrow\infty$ when $\tan\left(\frac{\omega L}{c}\right)\rightarrow\infty$ or when $\frac{\omega L}{c} = \frac{\pi}{2}, \frac{3\pi}{2}, \frac{5\pi}{2}, \ldots$ That is, resonance occurs when the applied frequency is

$$\omega = \frac{\pi}{2}\left(\frac{a}{L}\right), \quad \frac{3\pi}{2}\left(\frac{a}{L}\right), \quad \frac{5\pi}{2}\left(\frac{a}{L}\right), \ldots$$

6.4.4 Steady state response of a taut cable

A taut uniform cable of length L is fixed at both ends and subjected to a uniformly distributed load $F\cos\omega t$, as shown in Figure 6.21. The steady state response of the cable is required.

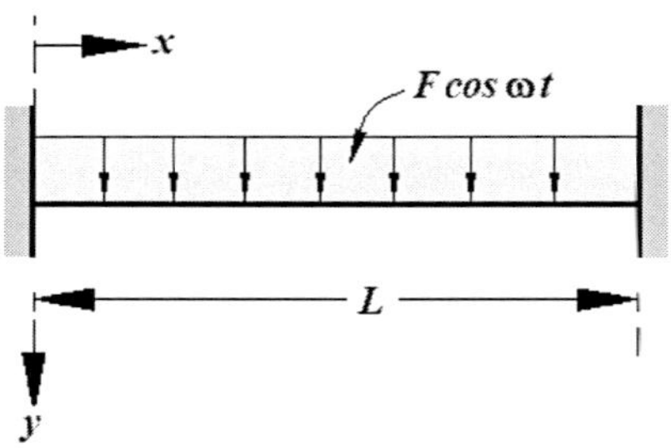

Figure 6.21: Fixed cable subjected to uniformly distributed load.

The equation of motion for free vibration of uniform cable, according to Equation (6.1), is

$$\mu\frac{\partial^2 v(x,t)}{\partial t^2} = T\frac{\partial^2 v(x,t)}{\partial x^2}$$

where $v(x,t)$ is the deflection along the y-axis.

With reference to the coordinate system and the loading direction in Figure 6.21 as well as by making use of the steps in the derivation of Equation (6.1), the equation for the forced vibration of the given cable, therefore, becomes

$$\mu \frac{\partial^2 v(x,t)}{\partial t^2} = T \frac{\partial^2 v(x,t)}{\partial x^2} + F\cos\omega t. \tag{6.81}$$

Since the forcing function is $F\cos\omega t$ and the system is linear so one can assume the general equation for the steady state vibration as $v(x,t) = V(x)\cos\omega t$. Thus,

$$\frac{\partial^2 v(x,t)}{\partial t^2} = -\omega^2 V(x)\cos\omega t, \qquad \frac{\partial^2 v(x,t)}{\partial x^2} = \frac{d^2 V(x)}{dx^2}\cos\omega t.$$

Substituting the above results into Equation (6.81), one has

$$\frac{d^2 V(x)}{dx^2} + \left(\frac{\omega}{c}\right)^2 V(x) = -\frac{F}{\mu c^2},$$

where $c = \sqrt{\frac{T}{\mu}}$ as defined by Equation (6.1). Therefore, the complete solution to this equation is

$$V(x) = A\cos\left(\frac{\omega}{c}x\right) + B\sin\left(\frac{\omega}{c}x\right) - \frac{F}{\mu\omega^2}, \tag{6.82}$$

where arbitrary constants, A and B are determined by the boundary conditions. Since the cable are fixed at both ends which means $V(0) = 0$ and $V(L) = 0$. The first boundary condition gives $A = \frac{F}{\mu\omega^2}$ and the second boundary condition results in

$$0 = \frac{F}{\mu\omega^2}\cos\left(\frac{\omega}{c}L\right) + B\sin\left(\frac{\omega}{c}L\right) - \frac{F}{\mu\omega^2}.$$

This leads to

$$B = \frac{F}{\mu\omega^2}\frac{\left[1 - \cos\left(\frac{\omega}{c}L\right)\right]}{\sin\left(\frac{\omega}{c}L\right)}.$$

Since $\tan\frac{u}{2} = \frac{1-\cos u}{\sin u}$, then $B = \frac{F}{\mu\omega^2}\tan\left(\frac{\omega}{2c}L\right)$. Therefore, the steady state vibration of the cable is

$$v(x,t) = \frac{F}{\mu\omega^2}\left[\cos\left(\frac{\omega}{c}x\right) + \tan\left(\frac{\omega}{2c}L\right)\sin\left(\frac{\omega}{c}x\right) - 1\right]\cos\omega t. \quad (6.83)$$

6.4.5 Response of uniform beam to harmonic force

Consider a simply supported damped uniform beam of cross-section area A and length L subjected to an applied force $P\sin\omega t$ at the quarter-span point (that is, $x = L/4$), as shown in Figure 6.22. Assuming the damping values $\varepsilon = \delta\omega = 0.05$ and for $\omega = \omega_1$, the steady-state response at mid-span of this beam is of interest here.

Applying Equation (6.65) for this system, one has the equation of motion as

$$\frac{d^2q_s(t)}{dt^2} + \delta\omega_s^2\frac{dq_s(t)}{dt} + \omega_s^2q_s(t) = \frac{W_s(L/4)\,P\sin\omega t}{\rho AL}, \quad (6.84)$$

where $\rho AL = \rho A\int_0^L[W_s(x)]^2dx$ and the mode shapes have been normalized. That is, $\int_0^L[W_s(x)]^2dx = L$. For completeness and demonstration, this normalization procedure will be included later. Note that $\rho AL = m$ is the mass of the beam.

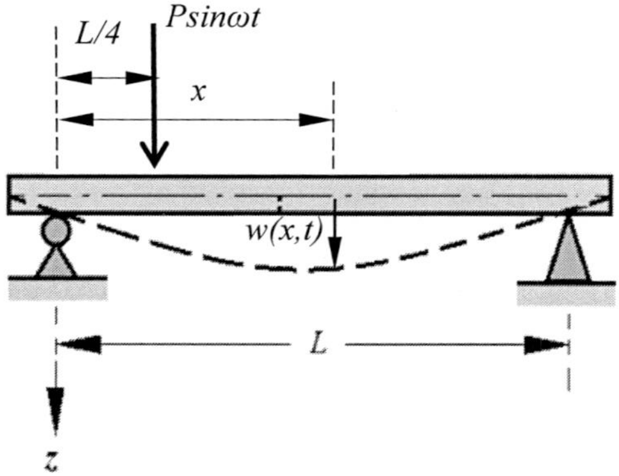

Figure 6.22: Simply supported damped uniform beam of cross-section area A and length L.

With reference to Equations (6.66) and (6.67), and by analogy to the response of a single dof system as presented in Chapter 3, Equation (6.84) gives

$$q_s(t) = \frac{W_s(L/4)\ P\sin(\omega t - \varphi_s)}{m\sqrt{(\omega_s^2 - \omega^2)^2 + (\varepsilon\omega_s^2)^2}},$$
$$\tan\varphi_s = \frac{\varepsilon\omega_s^2}{\omega_s^2 - \omega^2}. \tag{6.85}$$

Applying Equation (6.63), the steady-state response is

$$w(x,t) = \sum_s \frac{W_s(x)W_s(L/4)\ P\sin(\omega t - \varphi_s)}{m\sqrt{(\omega_s^2 - \omega^2)^2 + (\varepsilon\omega_s^2)^2}}. \tag{6.86}$$

According to the results in **Example 6.2**, the natural frequencies for a simply supported uniform beam are defined by

$$\omega_n = \beta_n^2\sqrt{\frac{EI}{\rho A}} = \left(\frac{n\pi}{L}\right)^2\sqrt{\frac{EI}{\rho A}}, \qquad n = 1, 2, 3, \ldots$$

Therefore,

$$\omega_s^2 = \left(\frac{s\pi}{L}\right)^4\frac{EI}{\rho A} = (s\pi)^4\frac{EI}{L^3 m}, \qquad s = 1, 2, 3, \ldots$$

Writing $\frac{\omega}{\omega_1} = r_1$, Equation (6.86) becomes

$$w(x,t) = \sum_s \frac{PL^3\ W_s(x)W_s(L/4)\ \sin(\omega t - \varphi_s)}{\pi^4 EI\sqrt{\left(s^4 - r_1^2\right)^2 + (\varepsilon s^4)^2}}. \tag{6.87}$$

According to the results in **Example 6.2**, the mode shapes for the simply supported beam are $W_s(x) = \alpha_n\sin\left(\frac{s\pi x}{L}\right)$, where α_n is the normalization constant so that

$$\int_0^L [W_s(x)]^2 dx = L.$$

In this particular problem, one has

$$\int_0^L \left[\alpha_n \sin\left(\frac{s\pi x}{L}\right)\right]^2 dx = \alpha_n^2 \int_0^L \left[\sin\left(\frac{s\pi x}{L}\right)\right]^2 dx$$

But $\int_0^L \sin\left(\frac{s\pi x}{L}\right)\sin\left(\frac{s\pi x}{L}\right) dx = \int_0^L \frac{1}{2}\left[1-\cos\left(\frac{2s\pi x}{L}\right)\right] dx = \frac{L}{2}$. Therefore, $\alpha_n^2 = 2$ giving $\alpha_n = \sqrt{2}$. That is, the normalized mode shapes are

$$W_s(x) = \sqrt{2}\sin\left(\frac{s\pi x}{L}\right).$$

Therefore, $W_s(L/4) = \sqrt{2}\sin\left(\frac{s\pi}{4}\right)$, which gives

$$W_1(L/4) = 1,\ \ W_2(L/4) = \sqrt{2},\ \text{and}\ W_3(L/4) = 1.$$

The dimensionless quantity $W_s^a = \frac{W_s(x)W_s(L/4)}{\sqrt{(s^4-r_1^2)^2+(\varepsilon s^4)^2}}$ at $x = \frac{L}{4}$ associated with the steady state response for the first three terms in Equation (6.87) for $r_1 = 1$ are presented in Table 6.3.

Table 6.3: Values of $W_s^a = \frac{W_s(x)W_s(L/4)}{\sqrt{(s^4-r_1^2)^2+(\varepsilon s^4)^2}}$ at $x = \frac{L}{4}$ for resonance number.

Resonance Number, s	1	2	3
W_s^a	20	0.09415	0.01250

Note that for $s > 3$ the dimensionless quantity W_s^a is much smaller than the first two terms for $s = 1, 2$ and therefore, numerically they are insignificant and not included in the present computation.

From Equation (6.85), $\varphi_1 = \frac{\pi}{2}$ rad, $\varphi_2 \approx 0.02$ rad, φ_s for $s \geq 3$ are approximately equal to zero.

Therefore, the steady state response, applying Equation (6.87) for the first three terms at $\omega = \omega_1$, is

$$w(x,t) = \left(\frac{PL^3}{\pi^4 EI}\right)\left[20\sin\left(\omega_1 t - \frac{\pi}{2}\right) + 0.09415\sin(\omega_1 t - 0.02) + 0.01250\sin(\omega_1 t)\right]. \tag{6.88}$$

The foregoing steps can be applied to evaluate the steady state response at different applied frequencies ω so that a plot for the relationship of $\frac{PL^3}{\pi^4 EI}$ and $\frac{\omega}{\omega_1}$ may be established.

6.4.6 Response of uniform beam to transient force

The tip deflection of a simply supported uniform undamped beam of cross-section area A and length L subjected at the mid-span to a transient force shown in Figure 6.23 is studied in this sub-section.

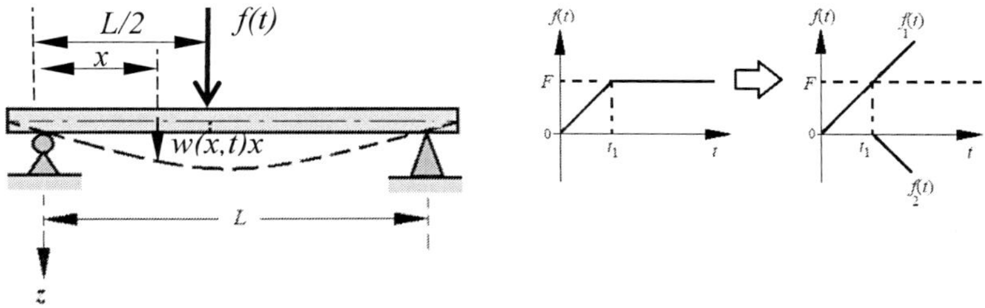

Figure 6.23: Simply supported beam under transient load at mid-span.

According to the results in **Example 6.2**, the natural frequencies for a simply supported uniform beam are defined by

$$\omega_n = \beta_n^2 \sqrt{\frac{EI}{\rho A}} = \left(\frac{n\pi}{L}\right)^2 \sqrt{\frac{EI}{\rho A}}, \qquad n = 1, 2, 3, \ldots$$

Therefore,

$$\omega_s^2 = \left(\frac{s\pi}{L}\right)^4 \frac{EI}{\rho A} = (s\pi)^4 \frac{EI}{L^3 m}, \qquad s = 1, 2, 3, \ldots$$

As presented in the last sub-section, the normalized mode shapes for the simply supported beam are

$$W_s(x) = \sqrt{2} \sin\left(\frac{s\pi x}{L}\right).$$

Therefore, $W_s(L/2) = \sqrt{2}\sin\left(\frac{s\pi}{2}\right)$, which gives

$$W_1(L/2) = \sqrt{2},\ \ W_2(L/2) = 0, \text{ and}$$
$$W_3(L/2) = -\sqrt{2}, \qquad W_4(L/2) = 0, \qquad W_5(L/2) = \sqrt{2}, \quad \ldots$$

The response at mid-span of the simply supported beam is

$$w(x,t) = \sum_s W_s(x) q_s(t) \tag{6.89}$$

where $q_s(t)$ is given by Equation (6.68) with $\delta = 0$ for the present case,

$$q_s(t) = \frac{Q_s f(t)}{\omega_s^2} - \frac{Q_s}{\omega_s^2}\int_0^t \frac{df(\tau)}{d\tau}\cos\omega_s(t-\tau)\,d\tau. \tag{6.90}$$

The transient excitation can be split into two parts as

$$f(t) = f_1(t) + f_2(t)$$

in which

$$f_1(t) = F\left(\frac{t}{t_1}\right), \quad \text{and} \quad f_2(t) = -F\left(\frac{t-t_1}{t_1}\right).$$

Substituting $f_1(t)$ into Equation (6.90), one has

$$q_s(t) = \frac{Q_s F\left(\frac{t}{t_1}\right)}{\omega_s^2} - \frac{Q_s}{\omega_s^2}\int_0^t F\left(\frac{1}{t_1}\right)\cos\omega_s(t-\tau)\,d\tau \quad \text{for } t \le t_1.$$

$$\begin{aligned} q_s(t) &= \frac{Q_s F t}{\omega_s^2 t_1} - \frac{Q_s F}{\omega_s^2 t_1}\int_0^t \cos\omega_s(t-\tau)\,d\tau \\ &= \frac{Q_s F t}{\omega_s^2 t_1} - \frac{Q_s F}{\omega_s^3 t_1}\left[-\sin\omega_s(t-\tau)\right]_0^t \\ &= \frac{Q_s F}{\omega_s^2 t_1}\left(t - \frac{\sin\omega_s t}{\omega_s}\right) \qquad \text{for } t \le t_1. \end{aligned} \tag{6.91a}$$

For the second ramp function beginning at t_1, the solution can be obtained by inspection of the above equation as

$$q_s(t) = -\frac{Q_s F}{\omega_s^2 \, t_1}\left[(t - t_1) - \frac{\sin \omega_s (t - t_1)}{\omega_s}\right] \qquad \text{for } t \geq t_1.$$

By superimposing the above two solutions, the response for $t \geq t_1$ is

$$q_s(t) = \frac{Q_s F}{\omega_s^2}\left[1 - \frac{\sin \omega_s t}{\omega_s \, t_1} + \frac{\sin \omega_s (t - t_1)}{\omega_s \, t_1}\right]. \tag{6.91b}$$

Substituting Equations (6.91) into Equation (6.89), one has the responses

$$w(x, t) = \sqrt{2}\left(\frac{F}{t_1}\right)\sum_s \frac{Q_s}{\omega_s^2} \sin\left(\frac{s\pi x}{L}\right)\left(t - \frac{\sin \omega_s t}{\omega_s}\right) \quad \text{for } t \leq t_1 \tag{6.92a}$$

and

$$w(x, t) = \sqrt{2}\, F \sum_s \frac{Q_s}{\omega_s^2} \sin\left(\frac{s\pi x}{L}\right)\left[1 - \frac{\sin \omega_s t}{\omega_s \, t_1} + \frac{\sin \omega_s (t - t_1)}{\omega_s \, t_1}\right] \quad \text{for } \; t \geq t_1. \tag{6.92b}$$

Clearly, at $t = t_1$ Equation (6.92*a*) and (6.92*b*) are identical.

Before leaving this sub-section it should be noted that when $s \geq 3$ the contributions of the higher order terms in the spatial domain to the responses given in Equation (6.92) is less than 1% with respect to the first term. Therefore, responses including the first three terms in Equation (6.92) are generally accepted as accurate.

6.4.7 Response of beam to uniformly distributed intensity

The steady state response of a uniformly, simply supported beam of cross-section area *A* and length *L* under a uniformly distributed forcing function of intensity $P \sin \omega t$, shown in Figure 6.24, is studied in this sub-section. The primary interest is the determination of the steady state response without the use of normal modes.

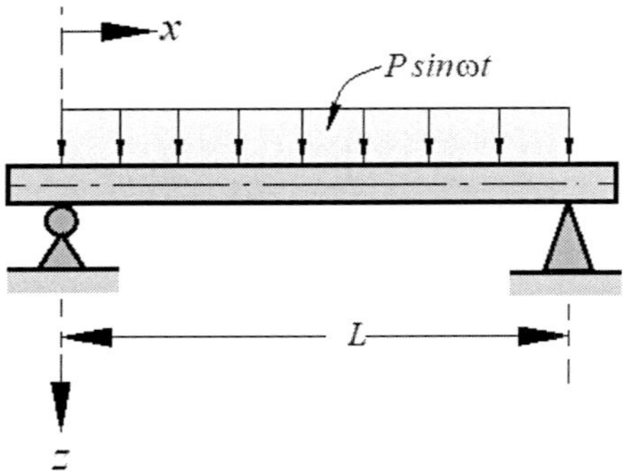

Figure 6.24: Simply supported uniform beam under distributed force.

With reference to Equation (6.21), the equation of motion for the present beam is

$$\rho A \frac{\partial^2 w}{\partial t^2} + EI \frac{\partial^4 w}{\partial x^4} = P \sin \omega t. \qquad (6.93)$$

Assume a solution to Equation (6.93) as

$$w(x,t) = W(x) \sin \omega t.$$

By making use of this expression and substituting into Equation (6.93), one obtains

$$\frac{d^4 W}{dx^4} - \beta^4 W = \frac{P}{EI}, \qquad (6.94)$$

where $\beta^4 = \left(\frac{\rho A}{EI}\right) \omega^2$ as defined in Equation (6.22). Thus,

$$W(x) = B_1 \sin \beta x + B_2 \cos \beta x + B_3 \sinh \beta x + B_4 \cosh \beta x - \frac{P}{EI\beta^4}$$

or

$$W(x) = B_1 \sin \beta x + B_2 \cos \beta x + B_3 \sinh \beta x + B_4 \cosh \beta x - \frac{P}{\rho A \omega^2}. \qquad (6.95)$$

For the present case, the boundary conditions are

$$W(0) = 0, \text{ and } \left.\frac{\mathrm{d}^2 W(x)}{\mathrm{d}x^2}\right|_{x=0} = 0, \; W(L) = 0, \text{ and } \left.\frac{\mathrm{d}^2 W(x)}{\mathrm{d}x^2}\right|_{x=L} = 0.$$

From the conditions on the lhs of the beam, one has

$$B_2 + B_4 = \frac{P}{\rho A \omega^2}, \qquad \beta^2(-B_2 + B_4) = 0$$

which give $B_2 = B_4 = \frac{P}{2\rho A\omega^2}$.

On the other hand, from the two conditions on the rhs of the beam, one obtains

$$0 = B_1 \sin\beta L + B_3 \sinh\beta L + \frac{P}{2\rho A\omega^2}(\cos\beta L + \cosh\beta L) - \frac{P}{\rho A\omega^2},$$

and

$$0 = \beta^2\left[- B_1 \sin\beta L + B_3 \sinh\beta L + \frac{P}{2\rho A\omega^2}(-\cos\beta L + \cosh\beta L)\right].$$

From these last two equations, one has

$$2B_3 \sinh\beta L + \frac{P}{\rho A\omega^2}(\cosh\beta L) - \frac{P}{\rho A\omega^2} = 0$$
$$2B_3 \sinh\beta L = \frac{P}{\rho A\omega^2}(1 - \cosh\beta L).$$

However,

$$\sinh\beta L = 2\sinh\left(\frac{\beta L}{2}\right)\cosh\left(\frac{\beta L}{2}\right), \; \cosh\beta L - 1 = 2\sinh^2\left(\frac{\beta L}{2}\right).$$

Substituting for these hyperbolic trigonometric identities, one obtains

$$4B_3 \sinh\left(\frac{\beta L}{2}\right)\cosh\left(\frac{\beta L}{2}\right) = \frac{-P}{\rho A\omega^2}\, 2\sinh^2\left(\frac{\beta L}{2}\right)$$

which gives $B_3 = -\frac{P}{2\rho A\omega^2}\tanh\left(\frac{\beta L}{2}\right)$.

Subtracting one from the other for the last two end conditions, one has

$$2B_1 \sin\beta L + \frac{P}{\rho A\omega^2}\cos\beta L \quad - \frac{P}{\rho A\omega^2} = 0$$

$$2B_1 \sin\beta L = \frac{P}{\rho A\omega^2}\,(1 - \cos\beta L\,).$$

But $\sin\beta L = 2\sin\left(\frac{\beta L}{2}\right)\cos\left(\frac{\beta L}{2}\right)$, $1 - \cos\beta L \, = 2\sin^2\left(\frac{\beta L}{2}\right)$. Therefore,

$$B_1 = \frac{P}{2\rho A\omega^2}\tan\left(\frac{\beta L}{2}\right).$$

Substituting all the four constants into Equation (6.95),

$$W(x) = \frac{P}{2\rho A\omega^2}\Big[\tan\left(\frac{\beta L}{2}\right)\sin\beta x + \cos\beta x \\ - \tanh\left(\frac{\beta L}{2}\right)\sinh\beta x \\ + \cosh\beta x - 2\ \Big]. \tag{6.96}$$

But

$$\tan\left(\frac{\beta L}{2}\right)\sin\beta x + \cos\beta x = \frac{\sin\left(\frac{\beta L}{2}\right)\sin\beta x + \cos\beta x\cos\left(\frac{\beta L}{2}\right)}{\cos\left(\frac{\beta L}{2}\right)}.$$

The numerator on the rhs is $\cos\beta\left(\frac{L}{2} - x\right)$. Therefore,

$$\tan\left(\frac{\beta L}{2}\right)\sin\beta x + \cos\beta x = \frac{\cos\beta\left(\frac{L}{2} - x\right)}{\cos\left(\frac{\beta L}{2}\right)}.$$

Similarly,

$$- \tanh\left(\frac{\beta L}{2}\right)\sinh\beta x + \cosh\beta x \\ = \frac{-\sinh\left(\frac{\beta L}{2}\right)\sinh\beta x + \cosh\beta x\cosh\left(\frac{\beta L}{2}\right)}{\cosh\left(\frac{\beta L}{2}\right)}.$$

The numerator on the rhs is $\cosh\beta\left(\frac{L}{2} - x\right)$. Therefore, one has

$$-\tanh\left(\frac{\beta L}{2}\right)\sinh\beta x+\cosh\beta x=\frac{\cosh\beta\left(\frac{L}{2}-x\right)}{\cosh\left(\frac{\beta L}{2}\right)}.$$

By making use of Equation (6.97), the steady state response becomes

$$w(x,t)=\frac{P}{\rho A\omega^2}\left[\frac{\cos\beta\left(\frac{L}{2}-x\right)}{2\cos\beta\left(\frac{L}{2}\right)}+\frac{\cosh\beta\left(\frac{L}{2}-x\right)}{2\cosh\beta\left(\frac{L}{2}\right)}-1\right]\sin\omega t. \quad (6.97)$$

6.4.8 Response to travelling point load

Lateral vibration of beam with a travelling point load is of interest in the investigation of car- or truck-bridge interaction. When the weight of the vehicle is small compared with that of bridge, the system may be idealized as an elastic simply supported beam under a moving force. The main assumptions are that the beam is uniform and simply supported, and the damping in the system is zero. The moving velocity of the force is constant. In this sub-section, the response of the beam to travelling point load is studied for the cases of constant and harmonic moving forces, respectively.

6.4.8.1 *The case of constant travelling point load*

Consider a uniform Euler beam with a constant travelling point force P, which is assumed to move across the uniform beam (bridge) from the lhs to the rhs with a constant velocity U, starting from the lhs support at time $=0$, as shown in Figure 6.25. The moving force acts at $x=Ut$ at time t and can be represented by $p(x,t)=P\delta_D(x-Ut)$, where δ_D is the Dirac delta function. Then the equation of motion for this system is given by

$$\rho A\frac{\partial^2 w}{\partial t^2}+EI\frac{\partial^4 w}{\partial x^4}=P\delta_D(x-Ut). \quad (6.98)$$

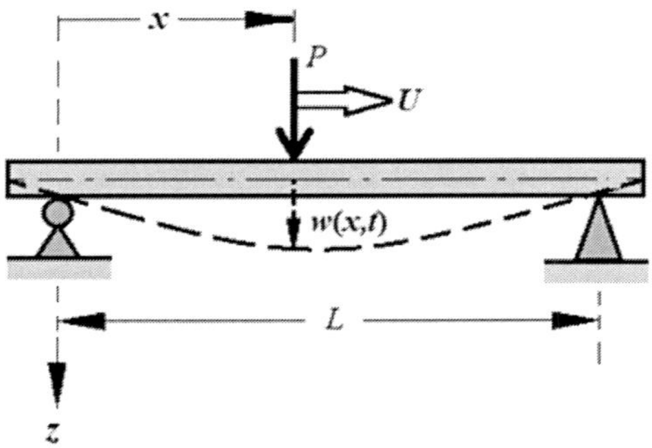

Figure 6.25: Euler beam with travelling point load.

Since the bridge is considered as a simply support beam, the response in terms of normal modes can be expressed by

$$\begin{aligned} w(x,t) &= \sum_s W_s(x) q_s(t) \\ &= \sum_s \sqrt{2}\sin\left(\frac{s\pi x}{L}\right) q_s(t) \end{aligned} \tag{6.99}$$

where the coefficient $\sqrt{2}$ is the normalization constant. Substituting Equation (6.99) into (6.98), we have

$$\frac{d^2 q_s(t)}{dt^2} + \omega_s^2 q_s(t) = \frac{\sqrt{2}}{m} P\sin\left(\frac{s\pi U t}{L}\right), \tag{6.100}$$

where $m = \rho AL$ and $\omega_s = \left(\frac{s\pi}{L}\right)^2 \sqrt{\frac{EI}{mL}}$ $(s = 1, 2, \dots)$ are the natural frequencies for simply support uniform beam.

If $\omega_s \neq \frac{s\pi U}{L}$, the complete solution to Equation (6.100) is given by

$$q_s(t) = B_1^s \sin \omega_s t + B_2^s \cos \omega_s t + \frac{\sqrt{2}}{m} P \frac{\sin\left(\frac{s\pi U t}{L}\right)}{\left[\omega_s^2 - \left(\frac{s\pi U}{L}\right)^2\right]}. \tag{6.101}$$

Assume that the beam starts from rest such that $q_s(0) = 0$ and $\left.\frac{dq_s(t)}{dt}\right|_{t=0} = 0$. It leads to that $B_1^s = -\left(\frac{\sqrt{2}}{m}\right) P \frac{\left(\frac{s\pi U}{L}\right)}{\left[\omega_s^2 - \left(\frac{s\pi U}{L}\right)^2\right]}$ and $B_2^s = 0$. Therefore, Equation (6.101) is reduced to

$$q_s(t) = \frac{\sqrt{2}}{m} P \left[\frac{\sin\left(\frac{s\pi U t}{L}\right) - \left(\frac{s\pi U}{L\omega_s}\right) \sin \omega_s t}{\omega_s^2 - \left(\frac{s\pi U}{L}\right)^2} \right]. \tag{6.102}$$

It should be noted that Equation (6.102) provides the response of the beam for $\omega_s \neq \frac{s\pi U}{L}$. As for $\omega_s = \frac{s\pi U}{L}$ $(s = 1,2,3, \dots \dots)$, Equation (6.100) becomes

$$\frac{d^2 q_s(t)}{dt^2} + \omega_s^2 q_s(t) = \frac{\sqrt{2}}{m} P \sin \omega_s t. \tag{6.103}$$

It is seen that the applied angular frequency on the rhs is equal to the natural frequency ω_s on the lhs in Equation (6.103). This means that the system is in resonance and the particular integral of the solution to Equation (6.103) grows with time when $\omega_s = \frac{s\pi U}{L}$. The complete solution to Equation (6.103) is given by

$$q_s(t) = B_1^s \sin \omega_s t + B_2^s \cos \omega_s t - \frac{Pt}{\sqrt{2}\, m} \left(\frac{\cos \omega_s t}{\omega_s} \right). \tag{6.104}$$

If $q_s(0) = 0$ and $\left.\frac{dq_s(t)}{dt}\right|_{t=0} = 0$, it leads to $B_1^s = \frac{P_0}{\sqrt{2}\, m\omega_s^2}$, $B_2^s = 0$, and then Equation (6.104) gives

$$q_s(t) = \frac{P}{\sqrt{2}\, m\omega_s^2} (\sin \omega_s t - \omega_s t \cos \omega_s t). \tag{6.105}$$

It is mentioned that the traveling force leaves the beam at $t = \frac{L}{U}$, so we need to consider the response of the beam in the time interval $[0, \frac{L}{U}]$.

Now, the special case of principal resonance when $\omega_1 = \frac{\pi U}{L}$ is concerned for application. The response of the beam is given by Equation (6.99), in which $q_1(t)$ is given by Equation (6.105) with $\omega_1 = \frac{\pi U}{L}$ and

$q_s(t)$ $(s > 1)$ is given by Equation (6.102) with $\omega_s = s^2\omega_1 = \frac{s^2\pi U}{L}$. In what follows, we only consider the response of the beam at time $t = \frac{L}{U} = \frac{\pi}{\omega_1}$ as an example.

At first, since $\omega_1 = \frac{\pi U}{L}$, Equation (6.105) gives

$$q_1(t) = \frac{P}{\sqrt{2}\, m\omega_1^2}\,(\sin\omega_1 t - \omega_1 t\cos\omega_1 t). \tag{6.106}$$

It can be shown that the maximum value of $q_1(t)$ occurs when $t = \frac{\pi}{\omega_1}$. Next, the numerator terms inside the square bracket of Equation (6.102) when $t = \frac{\pi}{\omega_1}$ becomes

$$\sin\left(\frac{s\pi Ut}{L}\right) - \left(\frac{s\pi U}{L\omega_s}\right)\sin\omega_s t = \sin\left(s\,\omega_1\frac{\pi}{\omega_1}\right) - \left(\frac{s\,\omega_1}{\omega_s}\right)\sin\left(\omega_s\frac{\pi}{\omega_1}\right),$$
$$= \sin(s\pi) - \left(\frac{1}{s}\right)\sin(s^2\pi) = 0.$$

It leads to that $q_s\left(\frac{\pi}{\omega_1}\right) = 0$ for $s > 1$, that is, the contribution of the higher modes ($s > 1$) to the displacement is zero when $t = \frac{L}{U} = \frac{\pi}{\omega_1}$. Therefore, according to Equation (6.99), the displacement of the beam when $t = \frac{\pi}{\omega_1}$ is given by

$$w\left(x, \frac{\pi}{\omega_1}\right) = \sum_s W_s(x) q_s\left(\frac{\pi}{\omega_1}\right) = \sqrt{2}\sin\left(\frac{\pi x}{L}\right) q_1\left(\frac{\pi}{\omega_1}\right). \tag{6.107}$$

Its maximum value occurs at $x = L/2$ and is given by

$$w_{max} = w\left(\frac{L}{2}, \frac{\pi}{\omega_1}\right) = \frac{P}{m\omega_1^2}\,(0 + \pi) = \frac{PL^3}{\pi^3 EI}. \tag{6.108}$$

Since the maximum static deflection of a beam for a force P at $x = \frac{L}{2}$ is $\frac{PL^3}{48EI}$, this means that w_{max} is 1.5481 times the maximum static deflection.

6.4.8.2 *The case of harmonic travelling point load*

Consider the case with a harmonic moving force $P\sin\omega t$, which is assumed to move across a simply support uniform beam from the lhs to the rhs with a constant velocity U, starting from the lhs support at time $t = 0$. Then the equation of motion for this system is given by

$$\rho A\frac{\partial^2 w}{\partial t^2} + EI\frac{\partial^4 w}{\partial x^4} = P\sin\omega t\,\delta_D(x - Ut), \tag{6.109}$$

the response in terms of normal modes is given by

$$\begin{aligned} w(x,t) &= \sum_s W_s(x)q_s(t) \\ &= \sum_s \sqrt{2}\sin\left(\frac{s\pi x}{L}\right)q_s(t) \end{aligned} \tag{6.110}$$

Substituting Equation (6.110) into Equation (6.109), we have

$$\frac{d^2 q_s(t)}{dt^2} + \omega_s^2 q_s(t) = \frac{\sqrt{2}P}{m}\sin\omega t\sin\left(\frac{s\pi Ut}{L}\right). \tag{6.111}$$

If $\omega_s \neq \omega \mp \frac{s\pi U}{L}$, the complete solution to Equation (6.111) is given by

$$q_s(t) = B_1^s\sin\omega_s t + B_2^s\cos\omega_s t + \frac{P}{\sqrt{2}\,m}\left[R_1^s(t) - R_2^s(t)\right], \tag{6.112}$$

in which B_1^s and B_2^s are constants to be determined by the initial conditions $q_s(0)$ and $\left.\frac{dq_s(t)}{dt}\right|_{t=0}$, that is,

$$R_1^s(t) = \frac{\cos\left(\omega - \frac{s\pi U}{L}\right)t}{\omega_s^2 - \left(\omega - \frac{s\pi U}{L}\right)^2}, \quad R_2^s(t) = \frac{\cos\left(\omega + \frac{s\pi U}{L}\right)t}{\omega_s^2 - \left(\omega + \frac{s\pi U}{L}\right)^2}.$$

If $\omega_s = \omega \mp \frac{s\pi U}{L}$ $(s = 1,2,3,\dots\dots)$, the system is in resonant states. Hence, large displacement of the beam is expected when $\omega = \omega_s \mp \frac{s\pi U}{L}$. Details about that are not discussed here due to the limitation of pages. It also should be noted that the responses of the beam from the above equations only apply to $t \leq \frac{L}{U}$, which means the moving load is on the beam.

6.4.9 Remarks

Since the systems considered in this chapter are linear, the forced responses of any combination of the above systems can be obtained by using the principle of superposition.

Appendix 6A

Bending Moment in Euler Beam

Consider Figure 6A.1 of the cross-section of a uniform Euler beam. With reference to this figure, the total displacement of the point P in the x-direction is

$$u = -z\alpha \tag{6A.1}$$

where the negative sign is due to the fact that section AB rotates cw in the figure. Note that there is no relative displacement of the point P along the z-direction because $z\cos\alpha = z$ since the angle α is small. Therefore, with the assumption that there is no other axial deformation along the axial direction, the strain along x-axis is

$$\varepsilon_{xx} = \frac{\partial u}{\partial x} = -z\frac{\partial^2 w}{\partial x^2}. \tag{6A.2}$$

Also, by definition, the modulus of elasticity is

$$E = \frac{\sigma_{xx}}{\varepsilon_{xx}}. \tag{6A.3}$$

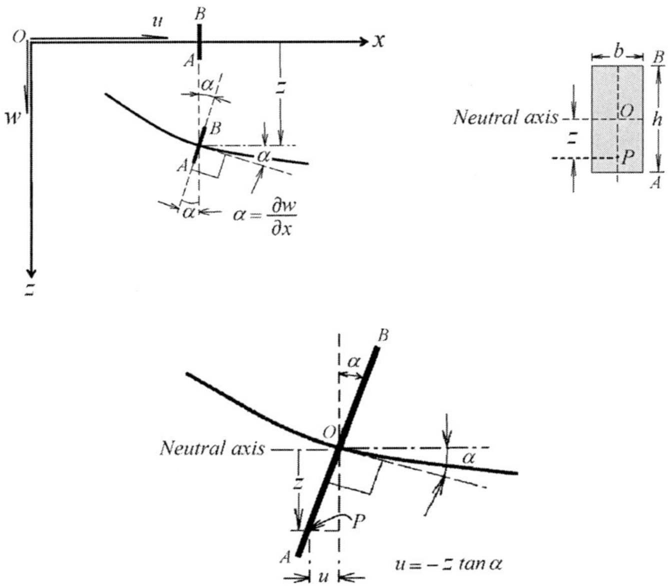

Figure 6A.1: Cross-section of beam in derivation of bending moment.

The stress σ_{xx} is related to the bending moment through the following equation

$$M = \int_{-\frac{h}{2}}^{\frac{h}{2}} \sigma_{xx}\,(bdz)z, \qquad (6A.4)$$

in which b and h are the breadth and depth of the cross-section area of the beam. Upon application of Equation (6A.2) through $(6A.4)$, one has

$$M = -\,E\frac{\partial^2 w}{\partial x^2}\int_{-\frac{h}{2}}^{\frac{h}{2}} bz^2\,dz \;=\; -\,EI\,\frac{\partial^2 w}{\partial x^2} \qquad (6A.5)$$

since $\int_{-\frac{h}{2}}^{\frac{h}{2}} bz^2\,dz = I$ is the second moment of the area of the cross-section of the beam. Note, in the foregoing the cross-sectional area of the beam is rectangular. However, other forms of cross-sectional areas can be similarly dealt with.

Appendix 6B

Orthogonality Conditions for Beams

While the following presentation is for uniform beams the procedure can similarly be applied to non-uniform beams. For free vibrations of uniform beams, Equation (6.21) reduces to

$$\rho A \frac{\partial^2 w}{\partial t^2} + EI \frac{\partial^4 w}{\partial x^4} = 0$$

which is rewritten as

$$EI \frac{\partial^4 w}{\partial x^4} = -\rho A \frac{\partial^2 w}{\partial t^2}. \qquad (6B.1)$$

Assuming $W_r(x)$ is a normal mode, so that the transverse response for one mode is

$$w(x,t) = W_r(x) \sin(\omega_r t + \varphi_r). \qquad (6B.2)$$

The equation governing the mode shape is therefore given by

$$EI \frac{d^4 W_r(x)}{dx^4} = \rho A \omega_r^2 W_r(x). \qquad (6B.3)$$

Multiplying Equation (6*B*.3) by another mode shape $W_s(x)$, and integrating over the length,

$$EI \int_0^L W_s(x) \frac{d^4 W_r(x)}{dx^4} dx = \rho A \omega_r^2 \int_0^L W_s(x)\, W_r(x) dx. \qquad (6B.4)$$

Integrating the lhs by parts, with $u = W_s(x)$, and $dv = \frac{d^4 W_r(x)}{dx^4} dx$ so that (for brevity the integration limits are disregarded at this stage)

$$\int u dv = uv - \int v du$$

where $v = \int dv = \int \frac{d^4 W_r(x)}{dx^4} dx = \frac{d^3 W_r(x)}{dx^3}$. Thus, Equation (6B.4) becomes

$$EI\left[\left|W_s(x)\frac{d^3W_r(x)}{dx^3}\right|_0^L - \int_0^L \frac{d^3W_r(x)}{dx^3}\frac{dW_s(x)}{dx}dx\right] = \rho A\omega_r^2 \int_0^L W_s(x)\, W_r(x)dx.$$

Integrating the second term on the lhs by parts, with $u = \frac{dW_s(x)}{dx}$, and $dv = \frac{d^3W_r(x)}{dx^3}dx$ such that $\int udv = uv - \int vdu$, it gives

$$EI\left\{\left|W_s(x)\frac{d^3W_r(x)}{dx^3}\right|_0^L - \left[\left|\frac{dW_s(x)}{dx}\frac{d^2W_r(x)}{dx^2}\right|_0^L - \int_0^L \frac{d^2W_r(x)}{dx^2}\frac{d^2W_s(x)}{dx^2}dx\right]\right\} = \rho A\omega_r^2 \int_0^L W_s(x)\, W_r(x)dx. \quad (6B.5)$$

For standard or simple end conditions, that is, any combination of free, simply-supported, and clamped ends at $x = 0$ and $x = L$, the first and second terms on the lhs of Equation (6*B*.5) are zero. Therefore, Equation (6*B*.5) reduces to

$$EI \int_0^L \frac{d^2W_r(x)}{dx^2}\frac{d^2W_s(x)}{dx^2}dx = \rho A\omega_r^2 \int_0^L W_s(x)\, W_r(x)dx. \quad (6B.6)$$

Now, one can consider Equation (6*B*.3) for mode shape $W_s(x)$. That is,

$$EI\frac{d^4W_s(x)}{dx^4} = \rho A\omega_s^2 W_s(x). \quad (6B.7)$$

Then multiplying Equation (6*B*.7) by another mode shape $W_r(x)$, and integrating over the length of the beam,

$$EI\int_0^L W_r(x)\frac{d^4W_s(x)}{dx^4}dx = \rho A\omega_s^2 \int_0^L W_r(x)\, W_s(x)dx. \quad (6B.8)$$

Integrating the lhs of Equation (6*B*.8) by parts twice as that for the lhs of Equation (6*B*.4), and for similar reasoning in regards to the standard end conditions, one has

$$EI \int_0^L \frac{d^2W_s(x)}{dx^2}\frac{d^2W_r(x)}{dx^2}dx = \rho A\omega_s^2 \int_0^L W_r(x)\, W_s(x)dx. \qquad (6B.9)$$

Subtracting Equation (6*B*.9) from (6*B*.6), one obtains

$$0 = \rho A(\omega_r^2 - \omega_s^2)\int_0^L W_s(x)\, W_r(x)dx. \qquad (6B.10)$$

For $\omega_r \neq \omega_s$, therefore,

$$\int_0^L W_s(x)\, W_r(x)dx = 0. \qquad (6B.11)$$

Applying this result in Equation (6*B*.6) or (6*B*.9), one has

$$\int_0^L \frac{d^2W_r(x)}{dx^2}\frac{d^2W_s(x)}{dx^2}dx = 0. \qquad (6B.12)$$

Applying Equation (6*B*.11) to (6*B*.4), one concludes that

$$\int_0^L W_s(x)\frac{d^4W_r(x)}{dx^4}dx = 0. \qquad (6B.13)$$

Equations $(6B.11), (6B.12),$ and $(6B.13)$ are true for $\omega_r \neq \omega_s$.

The general end conditions is illustrated in Figure 6.7 in which the translational and rotational springs k_L, and k_L^R, and a concentrated mass m_L are included where the subscript *L* denotes at the lhs (that is, $x = 0$) and the superscript *R* refers to the rotational spring, consideration of the boundary conditions (6.28*a*) and (6.28*b*), written in terms of mode shape $W_r(x)$, for $EI\left|\frac{d^3W_r(x)}{dx^3}\right|_0^L$ and $EI\left|\frac{d^2W_r(x)}{dx^2}\right|_0^L$ in Equation (6*B*.5), it results in

$$W_s(0)W_r(0)(k_L - m_L\omega_r^2) + k_L^R\left[\frac{dW_s(0)}{dx}\right]\left[\frac{dW_r(0)}{dx}\right]$$
$$+ EI\int_0^L \frac{d^2W_r(x)}{dx^2}\frac{d^2W_s(x)}{dx^2}dx = \rho A\omega_r^2\int_0^L W_s(x)\, W_r(x)dx. \qquad (6B.14)$$

Repeating the above steps for mode shape $W_s(x)$, and multiplying the resulting equation by mode shape $W_r(x)$, one obtains an equation similar to Equation $(6B.14)$,

$$W_r(0)W_s(0)(k_L - m_L\omega_s^2) + k_L^R\left[\frac{dW_r(0)}{dx}\right]\left[\frac{dW_s(0)}{dx}\right] + EI\int_0^L \frac{d^2W_s(x)}{dx^2}\frac{d^2W_r(x)}{dx^2}dx = \rho A\omega_s^2\int_0^L W_r(x)\,W_s(x)dx. \quad (6B.15)$$

Subtracting Equation $(6B.15)$ from$(6B.14)$, one has

$$W_s(0)W_r(0)m_L(\omega_r^2 - \omega_s^2) = \rho A(\omega_r^2 - \omega_s^2)\int_0^L W_r(x)\,W_s(x)dx.$$

Moving all terms to the same side, it gives

$$W_s(0)W_r(0)m_L(\omega_r^2 - \omega_s^2) + \rho A(\omega_s^2 - \omega_r^2)\int_0^L W_r(x)\,W_s(x)dx = 0.$$

For $\omega_r \neq \omega_s$, therefore, the orthogonality of the general end conditions, replacing Equation$(6B.11)$, is

$$m_L W_s(0)W_r(0) + \rho A\int_0^L W_r(x)\,W_s(x)dx = 0. \quad (6B.16)$$

From Equations $(6B.15)$ and $(6B.16)$, the general form of Equation $(6B.12)$ is

$$EI\int_0^L \frac{d^2W_r(x)}{dx^2}\frac{d^2W_s(x)}{dx^2}dx + k_L W_r(0)W_s(0) + k_L^R\left[\frac{dW_r(0)}{dx}\right]\left[\frac{dW_s(0)}{dx}\right] = 0. \quad (6B.17)$$

From Equations $(6B.4)$ and $(6B.16)$, the orthogonality condition $(6B.13)$ is replaced by

$$EI\int_0^L W_s(x)\frac{d^4W_r(x)}{dx^4}dx + m_L\omega_r^2 W_s(0)W_r(0) = 0. \quad (6B.18)$$

Before leaving this appendix it should be noted that the foregoing general end conditions are associated with a concentrated mass, a translational spring, and a rotational spring at the lhs of the beam. Similar equations for orthogonality conditions can also be obtained for the case at the rhs of the beam or for the case with concentrated masses, translational, and rotational springs at both ends of the beam.

References

Balachandran, B. and Magrab, E. B. 2004. Vibrations. Pacific Grove, CA: Brooks/Cole.

Cowper, G. R. 1966. "The shear coefficient in Timoshenko's beam theory". Journal of Applied Mechanics, 33(2), 335-340.

Kaladi, V. M. 1984. "Vibration analysis of piping systems with a steady mean flow". MSc. Thesis, University of Calgary.

Meirovitch, L. 1967. Analytical Methods in Vibrations, New York: The Macmillan Company.

Païdoussis, M. P. 2014. Fluid-Structure Interactions: Slender Structures and Axial Flow, Vol. 1, Second Edition. San Diego, CA: Academic Press.

Seto, W. W. 1964. Theory and Problems of Mechanical Vibrations. New York: McGraw-Hill.

Spiegel, M. R. 1968. Mathematical Handbook of Formulas and Tables. New York: McGraw-Hill.

Thomson, W. T. 1993. Theory of Vibration with Applications, Fourth Edition. Englewood Cliffs, New Jersey: Prentice-Hall.

Timoshenko, S. and Gere, J. M. 1961. Theory of Elastic Stability, Second Edition. New York: McGraw-Hill.

To, C. W. S. 1982. "Vibration of a cantilever beam with a base excitation and tip mass". Journal of Sound and Vibration, 83(4), 453.

Tong, Kin. N. 1960. Theory of Mechanical Vibration. New York, London: John Wiley & Sons, Inc.

Warburton, G. B. 1976. The Dynamical Behaviour of Structures, Second Edition. Oxford, England: Pergamon Press.

Exercise Questions

Q18. Set up the differential equations of motion and the boundary conditions for the following vibratory systems without damping:

(a) The lateral vibration of a flexible elastic string with length L is fixed at two ends and stretched to a tension force S initially. A coil spring with constant $K = S/3L$ connects the point $x = L/3$ to a stationary point at one side of the string.

(b) The lateral vibration of a uniform beam with length L is simply supported at $x = 0$ and $x = L/3$.

(c) The lateral vibration of a uniform beam with length L is fixed at the end $x = 0$ and simply supported at the other end $x = L$, carrying a concentrated mass at the mid-point $x = L/2$.

(d) The swinging of a heavy flexible chain with one end hung from the ceiling, while the other end is free and connected with a concentrated mass (Tong, 1960, 318).

Q19. A uniform beam with length L is fixed at the end $x = 0$ and connected to a fixed frame at the other end $x = L$ by means a linear spring, which exerts an axial force on the beam proportional to the end displacement. Set up the differential equations of motion and the boundary conditionsfor its longitudinal vibration. Determine the natural frequencies and the normal modes of the system.

Q20. A round bar with length L and is fixed at the end $x = 0$ and free at the end $x = L$. The outside diameter of the bar varies linearly from d (at $x =$ 0) to $2d$ (at $x = L$.). Set up the differential equations of motion and the boundary conditions for its lateral vibration. Determine the natural frequencies and the normal modes of the system.

Q21. A uniform beam with length L is simply supported at both ends. Suppose that the transverse displacement of one support at $x = 0$ is given by

$$u(t,0) = a\sin\omega t \text{ when } 0 \le t \le \frac{\pi}{\omega},$$

$$u(t,0) = 0 \text{ when } t > \frac{\pi}{\omega},$$

and the other support at $x = L$ is fixed. Set up the differential equations of motion and the boundary conditions for its lateral vibration when damping is neglected. Find the expressions for lateral displacement $u(t,x)$ at any point x on the beam for $0 \le t \le \frac{\pi}{\omega}$ and $t > \pi/\omega$, respectively. Furthermore, if both supports are subjected to the above displacement simultaneously, how about the new expressions for lateral displacement?

Q22. A uniform cantilever beam of length L is clamped at the end $x = 0$ and subjected to a transverse harmonic force $Q \sin \omega t$ at the tip $x = L$. Neglecting damping, if the mass of the beam is m, find the displacement at the tip. Moreover, if the damping in the beam material is small and the damping coefficient is μ, find the amplitude of vibrations at the tip when the excitation frequency is equal to the r-th natural frequency.

Q23. A 50 m long bridge is considered as a uniform simply supported beam with mass per unit length 20,000 kg/m and flexural rigidity 5 GN m^2. If the lateral vibration of the beam is excited by a transverse harmonic load 5000 sin 20 π t caused by a car moving across the bridge with a constant velocity U. If damping is neglected, determine the lateral displacement $u(t, x)$ at any point on the beam. When the system is in the resonance state with the lowest resonant frequency, find the value of U and the maximum displacement of the bridge.

INDEX